TEUBNER-TEXTE zur Mathematik   Band 134

L. Jentsch / F. Tröltzsch (Hrsg.)

Problems and Methods in Mathematical Physics

TEUBNER-TEXTE zur Mathematik

Herausgegeben von
Prof. Dr. Jochen Brüning, Augsburg
Prof. Dr. Herbert Gajewski, Berlin
Prof. Dr. Herbert Kurke, Berlin
Prof. Dr. Hans Triebel, Jena

Die Reihe soll ein Forum für Beiträge zu aktuellen Problemstellungen der Mathematik sein. Besonderes Anliegen ist die Veröffentlichung von Darstellungen unterschiedlicher methodischer Ansätze, die das Wechselspiel zwischen Theorie und Anwendungen sowie zwischen Lehre und Forschung reflektieren. Thematische Schwerpunkte sind Analysis, Geometrie und Algebra.

In den Texten sollen sich sowohl Lebendigkeit und Originalität von Spezialvorlesungen und Seminaren als auch Diskussionsergebnisse aus Arbeitsgruppen widerspiegeln.

TEUBNER-TEXTE erscheinen in deutscher oder englischer Sprache.

Problems and Methods in Mathematical Physics

Edited by
Prof. Dr. Lothar Jentsch
Prof. Dr. Fredi Tröltzsch

Technical University, Chemnitz–Zwickau

 Springer Fachmedien Wiesbaden GmbH 1994

Prof. Dr. Lothar Jentsch

Born in 1933 in Leipzig. Studied mathematics in Leipzig from 1953 to 1958. Received Dr. rer. nat. in 1966, Dr. sc. nat. in 1970 and Dr. rer. nat. habil. in 1991 from the University of Leipzig. Since 1975 Professor of Analysis (1992 Professor of Applied Mathematics) at the Technical University of Chemnitz–Zwickau.

Fields of interest: Mathematical methods in solid mechanics and integral equations.

Prof. Dr. Fredi Tröltzsch

Born in 1951 in Dennheritz. Studied mathematics in Freiberg from 1969 to 1973. Received Dr. rer. nat. in 1978, Dr. sc. nat. in 1982 and Dr. rer. nat. habil. in 1991 from the Freiberg Mining Academy. Since 1988 Professor of Analysis (1992 Professor of Applied Mathematics) at the Technical University of Chemnitz–Zwickau.

Fields of interest: Optimal control and optimization in Banach spaces.

Die Deutsche Bibliothek – CIP-Einheitsaufnahme

Problems and methods in mathematical physics /
ed. by Lothar Jentsch ; Fredi Tröltzsch. -
Stuttgart ; Leipzig : Teubner, 1994
  (Teubner-Texte zur Mathematik ; 134)

B. G. Teubner Verlagsgesellschaft Leipzig 1994

NE: Jentsch, Lothar [Hrsg.] ; GT

ISBN 978-3-322-85162-8   ISBN 978-3-322-85161-1 (eBook)
DOI 10.1007/978-3-322-85161-1

Umschlaggestaltung: E. Kretschmer, Leipzig

Preface

This Teubner-Text contains selected contributions to the "10th Conference on Problems and Methods in Mathematical Physics (10th TMP)", held at the Technical University of Chemnitz-Zwickau in Chemnitz, Germany, September 13 – 17, 1993. The main topics of the conference were

Scientific Computing in Solid and Fluid Mechanics

Problems with Singularities

Boundary Integral Methods

Multilevel Finite Element Methods

Free Boundary Value Problems and Optimal Control

Numerical Functional Analysis.

The conference brought together more than 170 experts from 17 countries, and the 18 plenary and invited lectures as well as the 120 short lectures were devoted to plenty of problems and latest results in applied mathematical research. Undoubtedly, the majority of these contributions would have deserved publication. However, this volume does not intend to reflect the whole scientific contents of the conference in the usual form of proceedings. It is rather aiming at giving a survey of some recent trends in fields covered by the topics of this meeting. The organizing committee hopes that this Teubner–Text reflects the main spirit of the 10th TMP.

The conference was sponsored by Sächsisches Staatsministerium für Wissenschaft und Kunst, Deutsche Forschungsgemeinschaft (DFG), Commission of the European Communities (Directorate General for Science, Research and

Development), Stiftungsfond IBM Deutschland, Sparkasse Chemnitz, Parsytec Chemnitz, and International Science Foundation. The scientific committee, the organizing committee and the Faculty of Mathematics of the Technical University of Chemnitz-Zwickau express their most sincere gratitude to these institutions. Thanks to them, we were enabled to invite many mathematicians from eastern countries, in particular from countries belonging to the former Soviet Union.

Moreover, we would like to express our sincere thanks to the 19 authors, who delivered excellent papers in camera–ready form. The editors are indebted to the collegues of the Chemnitz Faculty of Mathematics, Prof. A. Böttcher, Dr. H. Goldberg, Dipl.-Math. A. Unger and Mrs. G. Weise for technical assistence in preparing this volume.

Chemnitz, April 1994

L. Jentsch

F. Tröltzsch

Contents

8

On the Problem of Distributions Multiplication

Anatoliĭ Antonevich, Yakov Radyno

Abstract

The basic ideas related to the introduction of "new generalized functions" which admit everywhere defined multiplication are demonstrated in the report.

1 Introduction

The theory of distributions has led to the essential progress in a number of mathematical disciplines such as the theory of linear partial differential equations and linear mathematical physics. L.Schwartz has shown that it is impossible to introduce a multiplication of distributions [1]. This fact is an obstacle to the usage of distributions in the theory of nonlinear equations and the theory of generalized coefficients equations and in particular, one arrives at the impossibility to give a meaning to such objects as δ^2, $\delta'\delta$ (δ is the Dirac function) which are widely used, for example, in quantum field theory.

The problem proved to be very natural and to have a lot of applications. Therefore it has attracted attention at once after the creation of the distribution theory. There is a survey on this subject in [2], for instance. In the given report the basic ideas used while solving of the problem of distributions multiplication are presented.

Nontriviality of this problem can be shown in the following example. Let $u, v \in D'(\mathbf{R}), u_n, v_n \in C^\infty(\mathbf{R}), u_n \to u, v_n \to v$ in $D'(\mathbf{R})$. The sequence of products $u_n v_n$ may not converge in $D'(\mathbf{R})$, but it may have a limit, depending on the choice of the sequences u_n and v_n. For example, for any $a \in C^\infty(\mathbf{R})$ the sequence $u_n(x) = a(x)\sin(nx)$ converges to zero in $D'(\mathbf{R})$, but the sequence $u_n^2(x)$ converges to $a^2(x)/2$.

Classical approach is based on the specific choice of the sequences u_n and v_n. Thus it is possible to define a product for some pairs u and v. Another approach is based on the introduction of new objects instead of distributions which having the basic properties of distributions, and admit a well defined multiplication and form an associative algebra. Different variants of introduction of such objects have been given by B.Damyanov and Chr.Christov [3], V.K.Ivanov [4], J.Colombeau [5] and Yu.Egorov [2]. A general scheme of constructing of such

10

algebras was proposed by the authors of given report in [6]. The use of this method has made it possible to construct a number of new similar algebras with the given additional properties.

We remark that this scheme is a very general one and it allows to solve a related problem of everywhere defined convolution.

Some of the already known constructions, for example, one of the nonstandard expansion, proved to be special cases of our construction.

2 The statement of the problem and the general scheme of the solution

Let E' be some space of distributions and let A be a densely embedded in E' algebra consisting of infinitely differentiable functions. It is required to construct a new algebra G and a linear embedding $j : E' \to G$, such that A embedded as an algebra, i.e.

$$j(ab) = \gamma(a)\gamma(b), \ a, b \in A. \tag{1}$$

There is a certain delicacy in the statement of the problem. The Shwartz example shows that there are no embeddings of space $D'(\mathbf{R})$ in an algebra, satisfying the condition: $j(au) = \gamma(a)\gamma(u)$, $a \in A, u \in E'$ in the case when $E' = D'(\mathbf{R}), A = C^\infty(\mathbf{R})$. Therefore we have to require the preservation of multiplication only for the elements of the algebra of smooth functions.

The construction, presented below does not use the fact that the space E' consists of distributions and have a more general character.

The initial objects in our construction are:

a) a Hausdorf topological vector space E' (for the elements of which it is required to define a product);

b) a locally-convex algebra A, everywhere densely embedded in E';

c) a family of linear continuous operators $R_{\phi,\epsilon} : E' \to A$, where $\phi \in \Phi, \epsilon \in I$; Φ is a given set, I is a given set with a given filter F, and for each fixed ϕ $R_{\phi,\epsilon}(u) \to u$ with respect to F in the topology of E' for any $u \in E'$. For a fixed ϕ the family $R_{\phi,\epsilon}, \epsilon \in I$, is called a method of regularization of elements of E' by the elements of A and the condition c) is in fact a specification of some set of regularization methods.

Let $\tilde{G}$ be a set of all the mappings from $\Phi \times I$ into A. This set is an algebra and the space E' is linearly embedded into this algebra by means of the mapping $j_0 : u \to R_{\phi,\epsilon}(u)$. In some situation even such "trivial" embedding enables us to get valuable results. However this construction does not solve the problem posed above, since the equality (1) is not satisfied. Moreover, the algebra $\tilde{G}$ is too wide: in concrete problems different elements of this algebra describe one and the same object and they should be identified.

The construction goes as follows. In the algebra $\widetilde{G}$ we specify a certain subalgebra G_M and a certain ideal N in it. The quotient algebra G_M/N is the desired object. Thus the construction is reduced to the choice of a suitable pair (G_M, N).

Two elements f and g of $\widetilde{G}$ are said to be weakly equivalent if $f(\phi, \epsilon) - g(\phi, \epsilon) \to 0$ in E' with respect to F for any $\phi \in \Phi$. The set N_0 of all the elements weakly equivalent to zero is not even a subalgebra. Therefore the weak equivalence relation being natural in other problems does not solve this one.

Theorem 1. *Suppose that the subalgebra G_M and ideal N satisfy the following conditions:*

i) for any $u \in E'$ the element $R_{\phi,\epsilon}(u) \in G_M$;

ii) $N \subset N_0$;

iii) the elements of the form $R_{\phi,\epsilon}(ab) - R_{\phi\epsilon}(a)R_{\phi\epsilon}(b)$ belong to N for any $a, b \in A$.

Then the space E' is embedded in the quotient algebra $G = G_M/N$ and A is embedded as a subalgebra.

If G_M and N are invariant with respect to a certain linear continuous operator $L : A \to A$ than this operator L can be extended in a natural way onto the algebra G.

The condition (i) requires the ideal N to be rather small while the condition (ii) require the same ideal to be rather large. Therefore the existence of such ideals is not obvious. It's explicit description depends on the choice of the family of regularization methods.

The way of choice of the subalgebra G_M and the ideal N can be suggested by the topology in A. Let the topology on A be be defined by a system of multiplicative seminorms, i.e., seminorms $(p_\alpha), \alpha \in \Lambda$, satisfying the condition

$$p_\alpha(ab) \leq C_\alpha p_\alpha(a)p_\alpha(b).$$

The scalar function $p_\alpha(R_{\phi,\epsilon}(u)), u \in E'$, gives a typical growth of the elements of G_M with respect to ϵ and the functions $p_\alpha(R_{\phi,\epsilon}(ab) - R_{\phi,\epsilon}(a)R_{\phi,\epsilon}(b)), a, b \in A$, give a rate of descending of elements of N.

Let H_α be some algebra of scalar functions defined on $\Phi \times I$, generated by the functions of the form $p_\alpha(R_{\phi,\epsilon}(u)), u \in E'$ and let N_α be an algebra of scalar functions on $\Phi \times I$, generated by the functions of the form

$$p_\alpha(R_{\phi,\epsilon}(ab) - R_{\phi,\epsilon}(a)R_{\phi,\epsilon}(b)), a, b \in A.$$

The following theorem enables us to reduce the analysis to a rather simple case of scalar functions.

Theorem 2. *Let for every α the algebra N_α be an ideal in H_α. Then the sets*

$$G_N(A) = \{f \in \widetilde{G} : \forall \alpha \; \exists h \in H_\alpha, p_\alpha(f(\phi, \epsilon)) \le h(\phi, \epsilon)\},$$

$$N(A) = \{f \in \widetilde{G} : \forall \alpha \; \exists h \in N_\alpha, p_\alpha(f(\phi, \epsilon)) \le h(\phi, \epsilon)\}$$

satisfy the condition of Theorem 1 and the quotient algebra $G = G_M(A)/N(A)$ gives the solution of the problem.

3 Colombeau algebra

The first example of a pair (G_N, N) was constructed by Colombeau.
Let $E' = D'(\mathbf{R}), A = C^\infty(\mathbf{R})$. Denote by Φ a denumerable set

$$\Phi = \{\phi_1, \phi_2, \ldots\}, \;\; \phi_q \in D(\mathbf{R}), \tag{2}$$

$$\int \phi_q(x)dx = 1, \;\; \int x^j \phi_q(x)dx = 0, \;\; j = 1, 2, \ldots, q.$$

Each of the functions ϕ_q defines a regularization method by the formula

$$R_{q,\epsilon}(u) = u * \phi_{q,\epsilon} = u_{q,\epsilon}, \;\; \phi_{q,\epsilon}(x) = \frac{1}{\epsilon}\phi_q\left(\frac{x}{\epsilon}\right), \;\; \epsilon \in I = (0,1) \subset \mathbf{R}.$$

There exists a multiplicative system of seminorms on the algebra $C^\infty(\mathbf{R})$:

$$R_{k,n}(u) = \sum_{j=0}^{k} \max_{|x|\le n} \frac{u^{(j)}(x)}{k!}, \;\; k = 1, 2, \ldots, \;\; n = 1, 2, \ldots.$$

The functions $p_{k,n}(u_{q,\epsilon})$ are majorised by some power of $1/\epsilon$ when $\epsilon \to 0$. Therefore it is natural to define

$$G_M = \{f_{q,\epsilon}(x) : \forall k, n \; \exists m, c : p_{k,n}(f_{q,\epsilon}) \le \frac{c}{\epsilon^m}\}.$$

The functions of the form $p_{kn}((ab)_{q,\epsilon} - a_{q,\epsilon}b_{q,\epsilon})$, $a, b \in C^\infty(\mathbf{R})$ have a power rate of decreasing when $\epsilon \to 0$. For a fixed q the set of power decreasing function is not an ideal in the algebra of power growthing functions. Colombeau observed and used the fact, that the rate of decreasing of these functions increase with the increase of q. Therefore the set

$$N = \{f_{q,\epsilon}(x) : \forall k, n \; \exists q_0, C_q : p_{k,n}(f_{q,\epsilon}) \le C_q \epsilon^{q-q_0}\}$$

is a ideal in G_M and the conditions of Theorems 1 and 2 are satisfied. The quotient algebra $G = G_N/N$ is a modification of the Colombeau algebra of "new generalized functions".

Notice that the embedding of the space $D'(\mathbf{R})$ in G depends on the choice of a sequence (ϕ_q), but $C^\infty(\mathbf{R})$ appears to be canonically embedded for any embedding of this type. The general scheme and understanding of the patterns of choice of regularization methods have enabled us to construct a number of new generalized function algebras. In particular, the following algebras are already constructed: an algebra such that the ultradistributions are embedded in it: an algebra with the space $S'(\mathbf{R})$ embedded in it and with everywhere defined multiplication, convolution and invertible Fourier transform [7]; an algebra of periodical new generalized functions [8], an algebra constructed on the base of Sobolev spaces [9].

All the constructions described above can in essence be reduced to the fact that new generalized function appear to be a family of smooth functions depending on parameters. The functions belonging to the same class not only have the same limit, but they also have the same way of convergence to this limit. From this point of view we can tell, that a new generalized function memorize the way of tending to the limit, i.e., they possess a memory. This concept allow the authors of the report to call such objects "mnemofunctions" (greek-memories).

4   Mnemonumbers

If an algebra A contains the unit, then there exists a subalgebra $\widetilde{C}$ in the quotient algebra G_M/N , generated by the functions with the scalar values. We call this algebra a mnemonumber algebra and it's elements are called mnemonumbers. The use of mnemonumbers proved to be convenient for the presentation of different calculation results. It is related, in particular, to the fact that there exist infinitely large and infinitely small mnemonumbers. Let E' be the dual space of some locally convex space E, $< f, \psi >$ be the value of a functional f on $\psi \in E$.

A $\widetilde{C}$ -linear mapping between E and $\widetilde{C}$ we shall call a $\widetilde{C}$-functional.

Every mnemofunction has a corresponding $\widetilde{C}$-functional defined on E by the formula

$$< \widetilde{f}, \psi >=< f(\phi, \epsilon), \psi >, \ \psi \in E.$$

A mnemofunction f and element $u \in E'$ are said to be associated if for any $\phi \in \Phi < f(\phi, \epsilon), \psi > \to < u, \psi >, \ \psi \in E$ with respect to the filter F.

Observe that many different mnemofunctions can be associated with the element $u \in E'$ and that $\widetilde{C}$-functional $\widetilde{f}$ does not define mnemofunction f uniquely. At first sight it seems to be a disadvantage of the theory . However from the other hand this ambiguity is a new step which opens the way to solving a number of problems. For example, simulation of particle by means of Dirac δ-function is quite suitable in linear problems. But such construction proved to be too crude in the nonlinear case and some difficulties appear here. To obtain the results we need additional information about an object under consideration. Such information is the indication of mnemofunction (associated with δ-function) which similates the particles considered.

5   Examples

5.1. Let δ_Φ be an image of the δ-function under the embedding of $D'(\mathbf{R})$ in G by means of Φ of the form (2). Then $\delta_\Phi = \frac{1}{\epsilon}\phi_q\left(\frac{x}{\epsilon}\right)$. Let

$$a_q = \int \phi_q^2(x)dx \neq 0, \ \psi_q(x) = \frac{1}{a_q}\phi_q^2(x), \ \tilde{a} = (a_q) \in \tilde{\mathbf{C}}.$$

The later sequence ψ_q also defines the embedding of $D'(\mathbf{R})$ in G ($C^\infty(\mathbf{R})$ is embedded as a vector space, but not as an algebra). Let δ_ψ be the image of δ-function under this embedding. Then $\delta_\Phi^2 = \frac{\tilde{a}}{\epsilon}\delta_\psi$, i.e. δ^2 has the form of mnemofunction δ_ψ with an infinitely great coefficient $\frac{\tilde{a}}{\epsilon}$ from $\tilde{\mathbf{C}}$.

5.2. The multiplication of continuous functions does not coincide with the multiplication in G, when $D'(\mathbf{R})$ is embedded in G. It does not mean that an ordinary multiplication is rejected, but in G it is made more precise by the use of infinitely small values. For example, let $|x|_\Phi$ be an image of function $|x|$ under embedding Φ. Then $|x|_\Phi^2 = x^2$, but $|x|_\Phi^2 - x^2 = \epsilon^3 c\delta_n$, where $c \in \tilde{\mathbf{C}}$, i.e., the difference is the δ-function with infinitely small coefficient. The difference between the new theory and classical one can be clarified by the example of expression $(|x|^2 - x^2)\delta^3$. In the classical theory $|x|^2 - x^2 = 0$ and δ^3 is not defined. In the new one we have $(|x|_\Phi^2 - x^2)\delta_\Phi^3 = c\delta_n$, where c is some mnemonumber. Calculation of $\tilde{\mathbf{C}}$-functional associated with the mnemofunctions can be reduced to the analysis of the asymptotic behavior of integrals, depending on parameter. Of course, such calculations have been made long before the creation of the theory of new generalized functions. For example, the asymptotic behavior of δ-shaped family $\frac{1}{\pi}\frac{1}{x^2+\epsilon^2}$, is analyzed in [11], i.e. the corresponding $\tilde{\mathbf{C}}$ -functional is evaluated.

Similar calculations are used in the theory of equations with a small parameter, the theory of singular perturbations, etc. The situation formed after the creation of the theory of new generalized functions is similar to that described in the mathematical "folklore". Young writes that after the appearance of Schwartz's book on the distributions theory mathematicians divided into 3 classes: first said that they know it better then Schwartz, second said that they have known it before Schwartz, and the reviewer of the book occured to get into the third class and to write that all of it was nonsense [12].

6 Differential equations in the space of mnemofunctions

Let us consider as a model the differential equation

$$u' = \delta u + \delta, \ u(-1) = 1. \tag{3}$$

This equation has no solutions in the space of continuous at zero functions. If the function u is discontinuous , then the product δu is not defined. It means that there is no concept of the solution of this equation in the classical theory. To consider the equation (3) in the space of mnemofunctions it is necessary to choose some mnemofunctions among those, associated with the δ-function, for example δ_Φ and δ_Ψ, and to analyze the equation

$$u' = \delta_\phi u + \delta_\psi. \tag{4}$$

The solution of this equation is associated with an ordinary function of the form

$$u_0(x) = \begin{cases} 1, & x \le 0 \\ \mu, & x > 0 \end{cases},$$

where the number μ depends on the choice we have made. We emphasize that the information about the number μ has been obtained as the result of the choice of mnemofunction, i.e., as the result of the refinement of the statement of the problem.

The authors thank German Research Foundation for the financial support of their participation in 10 TMP conference.

References

[1] Schwartz L. Sur l'impossibilite' de la multiplication des distributions. C.R. Acad. sci. Paris. 1954, v.239, 874 - 848.

[2] Egorov Yu.Y. On the theory of generalized functions. Uspekhi Math. Nauk. 1990, v.45, n.5(275), 3 - 40.

16

[3] Christov Chr., Damianov B.P. Asymptotic Functions - a New Class of Generalized Functions. Bulgar. J.Phys. p.1, 1978, n.6, 543 - 556, p.II , 1979, n.1, 3 -23, p.III, 1979, n.3, 245 - 256, p.IV, 1979, n.4, 377 -397.

[4] Ivanov V.K. The multiplication of distributions and the regularization of divergent integrals. Izv. VUZ. Math. 1971, n.1, 41 - 49.

[5] Colombeau J.F. New generalized functions and multiplication of distributions. North Holland, 1989.

[6] Antonevich A.B., Radyno Ya.V. On a general method of constructing algebras of generalized functions. - Soviet. Math. Dokl. 1991, v.43, n.3, 680 - 684.

[7] Radyno Ya.V., Ngo Fu Thchan, Sabra R. The Fourier transform in the algebra of new generalized functions. Dokl. RAN. 1992, v.327, n.1, 20 - 24.

[8] Mazel M., Radyno N. The algebras of new generalized periodical functions. Dokl. AN Belarus. 1992, v.36, n.78, 585 - 588.

[9] Antonevich A.B., Foorsenko N.V. On some model elliptic equation with δ-shaped potential. Differentsialnye uravneniya. 1994, v.30 (to appear).

[10] Antosik P., Micusinski J., Sikorski R. Theory of distributions. The sequential approach. Amsterdam-Warszawa, 1973.

[11] Tikhonov A.N., Samarskiĭ A.A. An asymptotic expansion of integrals with slowly descending kernel. Dokl. AN USSR. 1959, v.126, n.1, 26 - 29.

[12] Young L. Lectures on the calculus of variations and optimal control theory. W.B. Saunders company, 1969.

Anatoliĭ Antonevich, Yakov Radyno,
Belorussian State University,
Department of Mathematics,
av. Skoryny 4,
220050 Minsk,
Belarus,
e-mail: anton%mmf.bsu.minsk.by@relay.ussr.eu.net

Developments in Boundary Element Methods for Time-Dependent Problems

Martin Costabel

1 Introduction

Time dependence is an essential feature in many engineering applications that are modelled by partial differential equations and, eventually, by boundary integral equations. Boundary element methods (BEM) have been successfully applied to many such problems from fields like elastodynamics, fluid dynamics, or acoustics.

Such problems are frequently modelled by *hyperbolic* equations, and *elliptic* equations are obtained in the limit cases of stationary or time-harmonic problems or with the help of the Laplace transform. *Parabolic* problems can be obtained in their own right in problems of heat transfer or other diffusion problems or also as certain (large damping) limit cases: In electrodynamics, for instance, for small frequencies and large conductivities, the limiting parabolic equation is used to describe eddy current problems.

It has been known for a long time that parabolic and hyperbolic initial-boundary value problems allow a reduction to boundary integral equations in a way very similar to elliptic boundary value problems. But compared to the vast literature about the boundary integral equations for the elliptic case, covering both their theoretical properties and their numerical approximation, the mathematical treatment of the boundary integral equations for parabolic and hyperbolic problems is rather modest in volume. There have been, however, some significant developments in this area in recent years. In this article, we will try to give a survey over some of the main mathematical ideas involved. Given the fast growth of the field of BEM, both on the side of engineering applications and on the side of mathematical and numerical analysis, the list of references cannot, of course, be comprehensive.

2 Some highlights of the mathematical history of time-dependent BEM

In the wake of the big success of BEM for elliptic boundary value problems, there have been many experiences with time-dependent BEM from the engineering side at least since 1980. The books [2], [6] contain reports about such experiments for hyperbolic problems, and [6], [7], [23] for parabolic problems. We shall concentrate here on the description of *mathematical* ideas (while keeping in mind the needs of the applications).

In the field of boundary integral equations for *parabolic* problems, the classical subject is, just as for elliptic problems, the study of integral equations of the second kind. These are of Volterra type (and for more than one space dimension, of Fredholm-Volterra type), and their theory for the case of *smooth* boundaries is summarized in POGORZELSKI's works [26]. This theory works in L^p spaces and has as its basic result the fact that such a Fredholm-Volterra integral operator K has the typical Volterra property of being quasi-nilpotent, i. e., $\lambda + K$ is invertible for any $\lambda \neq 0$.

Recent generalizations of this classical theory have the common feature that they closely (and sometimes amazingly closely) parallel similar developments for the case of elliptic equations.

Thus PIRIOU's parabolic pseudodifferential operators [24, 25] treat parabolic equations as anisotropic elliptic equations. PIRIOU's calculus contains the complete theory (regularity and solvability in Sobolev spaces) for all the boundary integral operators arising in the treatment of the standard initial-boundary value problems for the heat equation, provided that the boundary is a closed smooth surface.

For closed non-smooth (Lipschitz) boundaries, the theory as well as a certain Nyström approximation method for the second-kind integral equations in L^p spaces was studied by DAHLBERG&VERCHOTA, BROWN, TORRES and others (see [1, 8, 13]). A collocation method for second-kind integral equations on certain non-smooth domains was studied in [12].

For elliptic problems, the *variational formulation* of boundary integral equations was the key point that allowed to import ideas from the field of finite element methods (FEM) into BEM. Thus the coercivity of the operator of the single layer potential, shown by NEDELEC&PLANCHARD [21] and HSIAO&WENDLAND [17] was essential in several respects: for the proof of convergence of *any* Galerkin approximation method (including recently studied methods based on modern ideas like multigrid, domain decomposition, h-p or adaptive methods); for the possibility to allow open boundaries arising in crack, screen, or antenna problems; and also for the analysis of the coupling of boundary integral methods with other variational methods, e. g. variational inequalities or FEM–BEM coupling.

The coercivity of the single layer *heat* potential had been conjectured for some time, but when the correct result (see Theorem 5.2) was found in 1987 simultaneously by D. ARNOLD and the author, its essentially "elliptic" nature came as somewhat of a surprise. A proof of this result and its application to Galerkin methods on smooth surfaces was published by ARNOLD&NOON [3, 22]. Different methods of proof, generalizations of the result to other integral operators constituting the Calderón projector for the heat equation and to non-smooth surfaces as well as applications to various initial-boundary value problems and to the coupling with FEM were given by the author [11]. A generalization to another parabolic problem describing the Stokes flow, was given by HEBEKER&HSIAO [16].

A new and different development is LUBICH's "operational quadrature method" which is not based on the analogy with elliptic problems but rather takes the evolutional nature of the problem seriously. It allows to combine numerical methods for ordinary differential equations with standard BEM for the spatial (elliptic) part. LUBICH&SCHNEIDER [20] applied this technique to space-time integral equations for the heat equation, and LUBICH [19] showed how to apply it even to hyperbolic problems.

In the field of boundary integral equations for *hyperbolic problems*, apart from LUBICH's work just mentioned, variational methods are dominating, too. Here no (classical or recent) theory of second-kind integral equations seems to be available. BAMBERGER&HA DUONG [4] used a variational formulation of wave equation problems by J. L. Lions and local Fourier analysis to show coercivity of the single layer potential for the wave equation. Here the non-elliptic nature of the problem is felt in the loss of some regularity and also in the exponential growth of the stability constants with respect to time. This technique has been generalized to elastodynamics [5, 9] and to electrodynamics [27, 14, 28]. An independent development of variational methods for the space-time boundary integral equations in elastodynamics is described by KHUTORYANSKY [18, 29].

3   Notations

We will now study some of the above-mentioned ideas in closer detail. We consider only the simplest model problem of each type. Let $\Omega \subset \mathbb{R}^n$, $(n \geq 2)$, be a domain with compact boundary Γ. The outer normal derivative is denoted by ∂_n. Let $T > 0$ be fixed. We denote by Q the space-time cylinder over Ω and Σ its lateral boundary:

$$Q = (0,T) \times \Omega ; \quad \Sigma = (0,T) \times \Gamma ; \quad \partial Q = (\{0\} \times \overline{\Omega}) \cup \Sigma \cup (\{T\} \times \overline{\Omega}).$$

Elliptic problem (with frequency $\omega \in \mathbb{C}$):

$$\begin{aligned}
&(\Delta + \omega^2)u = 0 \quad \text{in } \Omega ; \\
&u = g \ \text{(Dirichlet)} \quad \text{or } \partial_n u = h \ \text{(Neumann)} \quad \text{on } \Gamma ; \\
&\text{radiation condition at } \infty .
\end{aligned} \qquad (\mathscr{E})$$

Parabolic problem:

$$\begin{aligned}
&(\partial_t - \Delta)u = 0 \quad \text{in } Q ; \\
&u = g \ \text{(Dirichlet)} \quad \text{or } \partial_n u = h \ \text{(Neumann)} \quad \text{on } \Sigma ; \\
&u = 0 \quad \text{for } t \leq 0 .
\end{aligned} \qquad (\mathscr{P})$$

Hyperbolic problem:

$$\begin{aligned}
&(\partial_t^2 - \Delta)u = 0 \quad \text{in } Q ; \\
&u = g \ \text{(Dirichlet)} \quad \text{or } \partial_n u = h \ \text{(Neumann)} \quad \text{on } \Sigma ; \\
&u = 0 \quad \text{for } t \leq 0 .
\end{aligned} \qquad (\mathscr{H})$$

4 Representation formulas and integral operators

We derive boundary integral equations by a general method that is valid (under suitable hypotheses on the data, C^∞ will certainly suffice...) in the same way for all 3 types of problems. In fact, what counts for $(\mathscr{P})$ and $(\mathscr{H})$ is the fact that the boundary Σ is non-characteristic.

The first ingredient for a BEM is a fundamental solution G. In 3D we have, respectively:

$$G_\omega(x) = \frac{e^{i\omega|x|}}{4\pi|x|} \qquad (\mathscr{E})$$

$$G(t,x) = \begin{cases} (4\pi t)^{-3/2} e^{-\frac{|x|^2}{4t}} & (t \geq 0) \\ 0 & (t \leq 0) \end{cases} \qquad (\mathscr{P})$$

$$G(t,x) = \frac{1}{4\pi|x|}\, \delta(t - |x|) . \qquad (\mathscr{H})$$

From Green's formula, the following representation formulas follow for a solution u of the homogeneous partial differential equation and $x \notin \Gamma$; $[v]$ denotes the jump of v across Γ:

$$u(x) = \int_\Gamma \{\partial_{n(y)} G(x-y)[u(y)] - G(x-y)[\partial_n u(y)]\}\, d\sigma(y) \qquad (\mathscr{E})$$

$$u(t,x) = \int_0^t \!\!\int_\Gamma \{\partial_{n(y)} G(t-s, x-y)[u(s,y)] - G(t-s, x-y)[\partial_n u(y)]\}\, d\sigma(y) \quad (\mathscr{P})$$

$$u(t,x) = \int_0^t \!\!\int_\Gamma \{\partial_{n(y)} G(t-s, x-y)[u(s,y)] - G(t-s, x-y)[\partial_n u(y)]\}\, d\sigma(y) \quad (\mathscr{H})$$

$$= \int_\Gamma \Big\{ \partial_{n(y)} \frac{1}{4\pi|x-y|}[u(t-|x-y|, y)] - \frac{\partial_{n(y)}|x-y|}{4\pi|x-y|}[\partial_t u(t-|x-y|, y)]$$

$$- \frac{1}{4\pi|x-y|}[\partial_n u(t-|x-y|, y)] \Big\}\, d\sigma(y)\,.$$

Thus the representation in the parabolic case uses integration over the past portion of Σ, whereas in the hyperbolic case, only the intersection of the interior of the backward light cone with Σ is involved. In 3D, where Huyghens' principle is valid, the last formula shows that the integration can be restricted to Γ, giving a very simple representation by "retarded potentials".

We can write all 3 representation formulas in a unified way, thereby introducing the single layer potential $\mathscr{S}$ and the double layer potential $\mathscr{D}$:

$$u = \mathscr{D}([u]) - \mathscr{S}([\partial_n u])\,.$$

There hold the classical jump relations

$$
\begin{array}{llll}
[\mathscr{D}v] & = & v & ; \\
[\mathscr{S}\varphi] & = & 0 & ;
\end{array}
\qquad
\begin{array}{llll}
[\partial_n \mathscr{D}v] & = & 0 & ; \\
[\partial_n \mathscr{S}\varphi] & = & -\varphi & .
\end{array}
$$

It appears therefore natural to introduce the boundary operators from the one-sided traces on the exterior (Γ^+) and interior (Γ^-) of Γ:

$$
\begin{array}{lll}
V & := & \mathscr{S}\big|_\Gamma \\[4pt]
K & := & \tfrac{1}{2}\big(\mathscr{D}\big|_{\Gamma^+} + \mathscr{D}\big|_{\Gamma^-}\big) \\[4pt]
K' & := & \tfrac{1}{2}\big(\partial_n \mathscr{S}\big|_{\Gamma^+} + \partial_n \mathscr{S}\big|_{\Gamma^-}\big) \\[4pt]
W & := & -\partial_n \mathscr{D}\big|_\Gamma
\end{array}
\qquad
\begin{array}{l}
\text{(single layer potential)} \\[4pt]
\text{(double layer potential)} \\[4pt]
\text{(normal derivative of single layer potential)} \\[4pt]
\text{(normal derivative of double layer potential)}
\end{array}
$$

In the standard way, the jump relations together with these definitions lead to boundary integral equations for the Dirichlet and Neumann problems. Typically one has a choice of at least 4 equations for each problem: The first 2 equations come from taking the traces in the representation formula ("direct method"), the third one comes form a single layer representation

$$u = \mathscr{S}\psi \quad \text{with unknown } \psi$$

and the fourth one from a double layer representation

$$u = \mathscr{D}w \quad \text{with unknown } w :$$

For the exterior Dirichlet problem $(u\big|_\Gamma = g$ given, $\partial_n u\big|_\Gamma = \varphi$ unknown):

$$
\begin{array}{llll}
(D1) & V\varphi & = & (-\tfrac{1}{2} + K)g \\
(D2) & (\tfrac{1}{2} + K')\varphi & = & -Wg \\
(D3) & V\psi & = & g \\
(D4) & (\tfrac{1}{2} + K)w & = & g
\end{array}
$$

For the exterior Neumann problem $(u\big|_\Gamma = g = v$ unknown, $\partial_n u\big|_\Gamma = h$ given):

$$
\begin{array}{llll}
(N1) & (\tfrac{1}{2} - K)v & = & -Vh \\
(N2) & Wv & = & -(\tfrac{1}{2} + K')h \\
(N3) & (\tfrac{1}{2} - K')\psi & = & -h \\
(N4) & Ww & = & -h
\end{array}
$$

Remember that this formal derivation is rigorously valid for all 3 types of problems. One notes that second-kind and first-kind integral equations alternate nicely. For open surfaces, however, only the first-kind integral equations exist. The reason is that a boundary value problem on an open surface fixes not only a one-sided trace but also the jump of the solution; and therefore the representation formula coincides with a single layer representation for the Dirichlet problem and with a double layer potential representation for the Neumann problem.

For reasons mentioned above, we will not treat the second-kind boundary integral equations in detail here. Suffice it to say that the key observation in the parabolic case is the fact that for smooth Γ, the operator norm in $L^p(\Sigma)$ of the weakly singular operator K tends to 0 as $T \to 0$. This implies that $\tfrac{1}{2} \pm K$ and $\tfrac{1}{2} \pm K'$ are isomorphisms in L^p (and also in C^m), first for small T and then by iteration for all T. If Γ has corners, this argument breaks down, and quite different methods, including variational arguments, have to be used.

5 First-kind integral operators: Green's formula

In the following, we restrict the presentation to the single layer potential operator V. We emphasize, however, that a completely analogous theory is available for the hypersingular operator W.

The variational methods for the first-kind integral operators are based on the first Green formula which gives, together with the jump relations, a formula valid again for all 3 types of equations: If φ and ψ are given on Γ or Σ, satisfy a finite number of conditions guaranteeing the convergence of the integrals on the right hand side of (5.1) and

$$u = \mathscr{S}\varphi, \qquad v = \mathscr{S}\psi,$$

then

$$\int_\Gamma \varphi V\psi\, d\sigma = \int_{\mathbb{R}^n\backslash\Gamma} \{\nabla u \cdot \nabla v + u\Delta v\}\, dx. \tag{5.1}$$

5.1 $(\mathscr{E})$

For the elliptic case, we obtain ($<\cdot,\cdot>_\Gamma$ denotes L^2 duality on Γ);

$$<\varphi, V\varphi>_\Gamma = \int_{\mathbb{R}^n\backslash\Gamma} (|\nabla u|^2 - \omega^2|u|^2)\, dx.$$

This gives the following theorem (see [10]) that serves as a model for the other two types:

Theorem 5.1 *Let Γ be a bounded Lipschitz surface, open or closed. $H^{1/2}(\Gamma)$ and $H^{-1/2}(\Gamma)$ denote the usual Sobolev spaces, and $\tilde{H}^{-1/2}(\Gamma)$ for an open surface is the dual of $H^{1/2}(\Gamma)$. Then*
(i) For $\omega = 0$, $n \geq 3$: $V : \tilde{H}^{-1/2}(\Gamma) \to H^{1/2}(\Gamma)$ is an isomorphism, and there is an $\alpha > 0$ such that

$$<\varphi, V\varphi>_\Gamma \geq \alpha\|\varphi\|^2_{\tilde{H}^{-1/2}(\Gamma)}.$$

(ii) For any ω and n, there is an $\alpha > 0$ and a compact quadratic form k on $\tilde{H}^{-1/2}(\Gamma)$ such that

$$\mathrm{Re}<\varphi, V\varphi>_\Gamma \geq \alpha\|\varphi\|^2_{\tilde{H}^{-1/2}(\Gamma)} - k(\varphi).$$

(iii) If ω is not an interior or exterior eigenfrequency, then V is an isomorphism, and every Galerkin method in $\tilde{H}^{-1/2}(\Gamma)$ for the equation $V\psi = g$ converges.

24

5.2 $(\mathscr{P})$

For the parabolic case, integration over t gives

$$
\begin{aligned}
<\varphi, V\varphi>_\Sigma &= \int_0^T \int_{\mathbb{R}^n \setminus \Gamma} \{|\nabla_x u(t,x)|^2 + \partial_t u \overline{u}\}\, dx\, dt \\
&= \int\int |\nabla_x u(t,x)|^2\, dx\, dt + \frac{1}{2}\int_{\mathbb{R}^n} |u(T,x)|^2\, dx \, .
\end{aligned}
$$

The positivity of the quadratic form associated to the operator V is evident. What is less evident is the nature of the energy norm for V, however. It turns out [3, 11] that one has to consider anisotropic Sobolev spaces of the following form

$$
\tilde{H}_0^{r,s}(\Sigma) = L^2(0,T;\, \tilde{H}^r(\Gamma)) \cap H_0^s(0,T;\, L^2(\Gamma)) \, .
$$

The index 0 indicates that zero initial conditions at $t = 0$ are incorporated. The optional $\tilde{}$ means zero boundary values on the boundary of the (open) manifold Γ. One has the following theorem which is actually simpler than its elliptic counterpart.

Theorem 5.2 *Let Γ be a bounded Lipschitz surface, open or closed, $n \geq 2$.*
(i) $V : \tilde{H}_0^{-\frac{1}{2},-\frac{1}{4}}(\Sigma) \to H_0^{r,s}(\Sigma)$ is an isomorphism, and there is an $\alpha > 0$ such that

$$
<\varphi, V\varphi>_\Sigma \geq \alpha \|\varphi\|^2_{-\frac{1}{2},-\frac{1}{4}} \, .
$$

(ii) Every Galerkin method in $\tilde{H}_0^{-\frac{1}{2},-\frac{1}{4}}(\Sigma)$ for the equation $V\psi = g$ converges. The Galerkin matrices have positive definite symmetric part. Typical error estimates are of the form

$$
\|\varphi - \varphi_{h,k}\|_{-\frac{1}{2},-\frac{1}{4}} \leq C\,(h^{r+\frac{1}{2}} + k^{(r+\frac{1}{2})/2})\|\varphi\|_{r,\frac{r}{2}} \, ,
$$

if $\varphi_{h,k}$ is the Galerkin solution in a tensor product space of splines of mesh-size k in time and finite elements of mesh-size h in space.

5.3 $(\mathscr{H})$

For the wave equation, choosing $\varphi = \overline{\psi}$ in the Green formula (5.1) does not give a positive definite expression. Instead, one can choose $\varphi = \overline{\partial_t \psi}$. This corresponds to the usual procedure in the weak formulation of the wave equation, and it gives

$$
\begin{aligned}
<\partial_t \varphi, V\varphi>_\Sigma \; &= \; \int_0^T \int_{\mathbb{R}^n \setminus \Gamma} \{\partial_t \nabla_x \overline{u} \cdot \nabla_x u + \overline{u}\partial_t^2 u\} \, dx \, dt \\
&= \; \frac{1}{2} \int_{\mathbb{R}^n \setminus \Gamma} \{|\nabla_x u(T,x)|^2 + |\partial_t u(T,x)|^2\} \, dx \, .
\end{aligned}
$$

Once again, as in the elliptic case, this shows the close relation of the operator V with the total energy of the system. In order to obtain a norm ($H^1(Q)$) on the right hand side, one can integrate a second time over t. But in any case, here the bilinear form $<\partial_t \varphi, V\varphi>_\Sigma$ will not be bounded in the same norm where its real part is positive. So there will be a loss of regularity, and any error estimate has to use two different norms. No "natural" energy space for the operator V presents itself.

6 First-kind integral operators: Fourier analysis

A closer view of what is going on can be obtained using space-time Fourier transformation. For this, one has to assume that Γ is flat, i. e. a subset of $\mathbb{R}^{n-1}$. Then all the operators are convolutions and as such are represented by multiplication operators in Fourier space. If Γ is not flat but smooth, then the results for the flat case describe the principal part of the operators. To construct a complete analysis, one has to consider lower order terms coming from coordinate transformations and localizations. Whereas this is a well-known technique in the elliptic and parabolic cases, namely part of the calculus of pseudodifferential operators, it has so far prevented the construction of a completely satisfactory theory for the hyperbolic case.

We denote the dual variables to (t,x) by (ω, ξ), and x' and ξ' are the variables related to $\Gamma \subset \mathbb{R}^{n-1}$. It is then easily seen that the form of the single layer potential is

$$
\widehat{V\psi}(\xi') = \frac{1}{2}(|\xi'|^2 - \omega^2)^{-\frac{1}{2}}\hat{\psi}(\xi') \tag{$\mathscr{E}$}
$$

$$
\widehat{V\psi}(\omega, \xi') = \frac{1}{2}(|\xi'|^2 + i\omega)^{-\frac{1}{2}}\hat{\psi}(\omega, \xi') \tag{$\mathscr{P}$}
$$

$$
\widehat{V\psi}(\omega, \xi') = \frac{1}{2}(|\xi'|^2 - \omega^2)^{-\frac{1}{2}}\hat{\psi}(\omega, \xi') \tag{$\mathscr{H}$}
$$

Note that $(\mathscr{E})$ and $(\mathscr{H})$ differ only in the role of ω: For $(\mathscr{E})$ it is a fixed parameter, for $(\mathscr{H})$ it is one of the variables, and this is crucial in the application of Parseval's formula for $<\varphi, V\varphi>$.

6.1 $(\mathscr{E})$

For the elliptic case, the preceding formula implies Theorem 5.1: If $\omega = 0$, then the function $\frac{1}{2}|\xi'|^{-1}$ is positive and for large $|\xi'|$ equivalent to $(1 + |\xi'|^2)^{-1/2}$, the Fourier weight defining the Sobolev space $H^{-1/2}(\Gamma)$. If $\omega \neq 0$, then the principal part (as $|\xi'| \to \infty$) is still $\frac{1}{2}|\xi'|^{-1}$, so only a compact perturbation is added. There is an additional observation by HA DUONG [15]: If ω is real, then $\frac{1}{2}(|\xi'|^2 - \omega^2)^{-\frac{1}{2}}$ is either positive or imaginary, so its real part is positive except on the bounded set $|\xi'| \leq |\omega|$. This implies

Proposition 6.1 *Let* $\omega^2 > 0$, Γ *flat,* $\operatorname{supp}\varphi$ *compact. Then there is an* $\alpha(\omega) > 0$ *such that*

$$\operatorname{Re} <\varphi, V\varphi>_\Gamma \geq \alpha(\omega) \|\varphi\|^2_{\tilde{H}^{-1/2}}.$$

The work of transforming this estimate into error estimates for the BEM in the hyperbolic case still has to be done.

6.2 $(\mathscr{P})$

For the parabolic case, the symbol of the single layer potential,

$$\sigma_V(\omega, \xi') = \frac{1}{2}(|\xi'|^2 + i\omega)^{-\frac{1}{2}}$$

has again positive real part. In addition, it is sectorial:

$$|\arg \sigma_V(\omega, \xi')| \leq \frac{\pi}{4}.$$

This has the consequence that its real part and absolute value are equivalent (an "elliptic" situation):

$$C_1 \left||\xi'|^2 + i\omega\right|^{-\frac{1}{2}} \leq \operatorname{Re}\sigma_V(\omega, \xi') \leq C_2 \left||\xi'|^2 + i\omega\right|^{-\frac{1}{2}}.$$

In addition, for large $|\xi'|^2 + |\omega|$, this is equivalent to $((1 + |\xi'|^2) + |\omega|)^{-1/2}$, the Fourier weight defining the space $H^{-\frac{1}{2},-\frac{1}{4}}(\Sigma)$. This explains Theorem 5.2. It also shows clearly the difference to the heat operator $\partial_t - \Delta$ itself: The symbol of the latter is $|\xi|^2 + i\omega$, and the real part and absolute value of this function are not equivalent.

6.3 $(\mathcal{H})$

In the hyperbolic case, the symbol σ_V does not have positive real part. Instead, one has to multiply it by $i\bar{\omega}$ and to use a complex frequency $\omega = \omega_R + i\omega_I$ with ω_I fixed. Then one gets

$$\operatorname{Re}\left(i\bar{\omega}(|\xi'|^2 - \omega^2)^{\frac{1}{2}}\right) \geq \frac{\omega_I}{2}(|\xi'|^2 + |\omega|^2)$$

and similar estimates given by BAMBERGER&HA DUONG [4]. One introduces another class of anisotropic Sobolev spaces of the form

$$H^{s,r}(\mathbb{R} \times \Gamma) = \{u \mid u, \partial_t^r u \in H^s(\mathbb{R} \times \Gamma)\}$$

with the norm

$$\|u\|_{r,s,\omega_I} = \int\limits_{\operatorname{Im}\omega=\omega_I} \int_{\mathbb{R}^{n-1}} |\omega|^{2r}(|\xi'|^2 + |\omega|^2)^s |\hat{u}(\omega,\xi')|^2 \, d\xi' \, d\omega \,.$$

We give one example of a theorem obtained in this way.

Theorem 6.2 *Let Γ be bounded and smooth, $r,s \in \mathbb{R}$. Then*
(i) $V : \tilde{H}_0^{s,r+\frac{1}{2}}(\Sigma) \to H_0^{s+1,r}(\Sigma)$ and $V^{-1} : H^{s+1,r+1}(\Sigma) \to \tilde{H}_0^{s,r}(\Sigma)$
   are continuous.
(ii) Let $\omega_I > 0$ and the bilinear form $a(\varphi,\psi)$ be defined by

$$a(\varphi,\psi) = \int_0^\infty e^{-2\omega_I t} \int_\Gamma (V\varphi)(t,x)\,\overline{\partial_t\psi}(t,x)\,d\sigma(x)\,dt \,.$$

Then there is an $\alpha > 0$ such that

$$\operatorname{Re} a(\varphi,\varphi) \geq \alpha\,\omega_I\,\|\varphi\|^2_{-\frac{1}{2},0,\omega_I} \,.$$

(iii) The Galerkin matrices for the scheme: Find $\varphi_N \in X_N$ such that

$$a(\varphi_N,\psi) = \,<g,\partial_t\psi>_\Sigma \quad \forall\psi \in X_N$$

have positive definite hermitian part, and there is an error estimate

$$\|\varphi - \varphi_N\|_{-\frac{1}{2},0,\omega_I} \leq C\,\omega_I^{-\frac{1}{2}} \inf_{\psi\in X_N} \|\varphi - \psi\|_{-\frac{1}{2},1,\omega_I} \,.$$

7 A time-stepping method

The Galerkin methods for parabolic and hyperbolic problems studied in the previous sections are *global in time*: They use the boundary data for $0 \leq t \leq T$ to compute the unknowns on the boundary and hence the solution u also for $t \in [0, T]$ in one step. Thus, in general, the Volterra convolution structure (causality) of the differential and integral equations will be lost by discretization. Only in special cases, this structure is conserved on the discrete level. If, for example, for the single layer heat potential, the trial functions are piecewise constant in time, then the Galerkin matrix will have the form

$$
\begin{pmatrix}
B_0 & & & 0 \\
B_1 & \ddots & & \\
\vdots & \ddots & \ddots & \\
B_{n_t} & \cdots & B_1 & B_0
\end{pmatrix}
\tag{7.1}
$$

Here the blocks B_j are $n_x \times n_x$ matrices (n_x = number of degrees of freedom of the space discretization; n_t = number of time steps). Thus only B_0 has to be inverted, and in order to increase n_t by one, only one new matrix B_{n_t+1} has to be computed. If, however, the order of approximation in time is increased, then the matrix will have more and more blocks above the diagonal and therefore lose the discrete causal structure. For piecewise linear approximation in time with the usual hat function basis, for example, the matrix will have the form

$$
\begin{pmatrix}
B_0 & B_{-1} & & 0 \\
\vdots & \ddots & \ddots & \\
\vdots & \ddots & \ddots & B_{-1} \\
B_{n_t} & \cdots & \cdots & B_0
\end{pmatrix}
$$

This problem vanishes if one replaces the space-time Galerkin boundary element approximation by a different method, the recently developped "operational quadrature" method of LUBICH [19, 20]. We will not spoil the reader's pleasure of reading these two papers by describing their contents in detail; we shall rather give an indication of the basic idea.

In this method, one discretizes not the space-time operator with its kernel as given above in section 4, but rather its Laplace transform with respect to time. This kernel corresponds to an elliptic problem, and for the space discretization one can choose any suitable BEM for elliptic problems, for instance collocation instead of Galerkin methods. For the time discretization, one chooses a discretization scheme for ordinary differential equations. This scheme has to satisfy certain stability conditions. If it is an explicit linear multistep method, then the resulting matrix will be of the block triangular form 7.1, although the method can be of higher order in time.

More precisely, the operational quadrature method considers an operator-valued convolution operator

$$g \mapsto \int_0^t k(t - \tau)\, g(\tau)\, d\tau =: K(\partial_t)\, g\,.$$

The basic object is the Laplace-transformed kernel $K(s) = \int_0^\infty e^{-st} k(t)\, dt$. With a linear multistep method for $y' = f(t, y)$,

$$a_0 y_n + a_1 y_{n-1} + \cdots + a_k y_{n-k} = h\left(b_0 f_n + \cdots + b_k f_{n-k}\right)$$

and its characteristic function,

$$\delta(\zeta) = (a_0 + a_1 \zeta + \cdots + a_k \zeta^k)/(b_0 + b_1 \zeta + \cdots + b_k \zeta^k)\,,$$

one constructs the following approximation of $K(\partial_t)g$:

$$K(\partial_t^h)g(t) = \sum_{j \geq 0} \omega_j g(t - jh)\,,$$

where the ω_j are the Taylor coefficients of $K(\delta(\zeta)/h)$ at $\zeta = 0$:

$$K\left(\frac{\delta(\zeta)}{h}\right) = \sum_{j \geq 0} \omega_j \zeta^j\,.$$

This discretization has the following decisive properties:

$$K_1(\partial_t^h) \cdot K_2(\partial_t^h) = (K_1 \cdot K_2)(\partial_t^h)\,;$$

if $y_j = (K(\partial_t^h)g)(jh)$ and $Y(\zeta) = \sum y_j \zeta^j$, $G(\zeta) = \sum g_j \zeta^j$ then

$$Y(\zeta) = K\left(\frac{\delta(\zeta)}{h}\right) G(\zeta)\,.$$

In the application to time-dependent BEM, one thinks of K as the inverse of the space discretization. $K(\delta(\zeta)/h)$ is then an approximation of the inverse of the space-time integral operator in question. The coefficients are therefore the result of an elliptic BEM performed for a certain number of complex frequencies $\frac{\delta(\zeta)}{h}$, where ζ runs through the nodes of some quadrature rule on a small circle in the complex plane. This method promises to be very efficient, in particular since it allows to combine some of the recently developped sophisticated BEM for elliptic problems with well-known high order methods for ordinary differential equations and fast methods for the computation of Fourier coefficients. How it competes, for example, with very simple direct space-time methods using retarded potentials for the 3D wave equation, remains to be seen, however.

References

[1] Adolfssohn, V., Jawerth, B., Torres, R.: A boundary integral method for parabolic equations in nonsmooth domains. Preprint 1993, to appear in Comm. Pure Appl. Math.

[2] Antes, H.: *Anwendungen der Methode der Randelemente in der Elastodynamik.* Stuttgart: Teubner 1988.

[3] Arnold, D. N., Noon, P. J.: Coercivity of the single layer heat potential. *J. Comput. Math.* **7**, 100–104 (1989).

[4] Bamberger, A., Ha Duong, T.: Formulation variationnelle espace-temps pour le calcul par potentiel retardé d'une onde acoustique. *Math. Meth. Appl. Sci.* **8**, 405–435 and 598–608 (1986).

[5] Becache, E.: Résolution par une méthode d'équations intégrales d'un problème de diffraction d'ondes élastiques transitoires par une fissure. Thèse de doctorat, Université Paris VI 1991.

[6] Brebbia, C. A., Telles, J. C. F., Wrobel, L. C.: *Boundary Element Techniques.* Berlin: Springer-Verlag 1984.

[7] Brebbia, C. A., Wrobel, L. A.: The solution of parabolic problems using the dual reciprocity boundary element. In: *Advanced Boundary Element Methods* (T. A. Cruse, ed.), pp. 55–72. Berlin: Springer-Verlag 1988.

[8] Brown, R. M.: The method of layer potentials for the heat equation in Lipschitz cylinders. *Amer. J. Math.* **111**, 339–379 (1989).

[9] Chudinovich, I. Y.: The boundary equation method in the third initial boundary value problem of the theory of elasticity. *Math. Meth. Appl. Sci.* **16**, 203–227 (1993).

[10] Costabel, M.: Boundary integral operators on Lipschitz domains: Elementary results. *SIAM J. Math. Anal.* **19**, 613–626 (1988).

[11] Costabel, M.: Boundary integral operators for the heat equation. *Integral Equations Oper. Theory* **13**, 498–552 (1990).

[12] Costabel, M., Onishi, K., Wendland, W. L.: A boundary element collocation method for the Neumann problem of the heat equation. In: *Inverse and Ill-posed Problems* (H. W. Engl, C. W. Groetsch, eds.), pp. 369–384. Academic Press 1987.

[13] Dahlberg, B., Verchota, G.: Galerkin methods for the boundary integral equations of elliptic equations in non-smooth domains. *Contemporary Mathematics* **107**, 39–60 (1990).

[14] Däschle, C.: Eine Raum-Zeit-Variationsformulierung zur Bestimmung der Potentiale für das elektromagnetische Streuproblem im Außenraum. Dissertation, Universität Freiburg 1992.

[15] Duong, T. H.: On the transient acoustic scattering by a flat object. Rapport Interne 194, CMAP, Ecole Polytechnique 1989.

[16] Hebeker, F. K., Hsiao, G. C.: On a boundary integral equation approach to a nonstationary problem of isothermal viscous compressible flows. Preprint 1988, to appear.

[17] Hsiao, G. C., Wendland, W. L.: A finite element method for some integral equations of the first kind. *J. Math. Anal. Appl.* **58**, 449–481 (1977).

[18] Хуторянский, Н. М.: Дискретные уравнения с положительно определенными и симметричными матрицами в методе граничных элементов для основных краебых задач теорий упругосты. Прикладные проблемы прочности и пластичности. Автомат. научных иссл. по прочности. Бсесоюз. межвуз. сб./Горк. ун-т. pp. 35–40 (1986).

[19] Lubich, C.: On the multistep time discretization of linear initial-boundary value problems and their boundary integral equations. *Numer. Math.* **67**, 365–390 (1994).

[20] Lubich, C., Schneider, R.: Time discretization of parabolic boundary integral equations. *Numer. Math.* **63**, 455–481 (1992).

[21] Nedelec, J.-C., Planchard, J.: Une méthode variationnelle d'éléments finis pour la résolution numérique d'un problème extérieur dans $\mathbb{R}^3$. *RAIRO* **7**, 105–129 (1973).

[22] Noon, P. J.: The single layer heat potential and Galerkin boundary element methods for the heat equation. Thesis, University of Maryland 1988.

[23] Onishi, K.: Galerkin method for boundary integral equations in transient heat conduction. In: *Boundary Elements IX* (C. A. Brebbia, W. L. Wendland, G. Kuhn, eds.), vol. 3, pp. 231–248. Berlin: Springer-Verlag 1987.

[24] Piriou, A.: Une classe d'opérateurs pseudo-différentiels du type de Volterra. *Ann. Inst. Fourier Grenoble* **20,1**, 77–94 (1970).

[25] Piriou, A.: Problèmes aux limites généraux pour des opérateurs differentiels paraboliques dans un domaine borné. *Ann. Inst. Fourier Grenoble* **21,1**, 59–78 (1971).

[26] Pogorzelski, W.: *Integral Equations and their Applications*. Oxford: Pergamon Press 1966.

[27] Pujols, A.: Equations intégrales espace-temps pour le système de Maxwell – application au calcul de la surface équivalente Radar. Thèse de doctorat, Université Bordeaux I 1991.

[28] Terrasse, I.: Résolution mathématique et numérique des équations de Maxwell instationnaires par une méthode de potentiels retardés. Thèse de doctorat, Ecole Polytechnique 1993.

[29] Угодчиков, А. Г., Хуторянский, Н. М.: *Метод Граничных Элементов в Механике Деформируемого Твердого Тела*. Казань: Издательство Казанского Университета 1986.

Martin Costabel
Institut Mathématique
Université de Rennes 1
Campus de Beaulieu
F-35042 Rennes Cedex
e-mail: costabel@univ-rennes1.fr

Finite Element Variational Crimes in the Solution of Nonlinear Stationary Problems

Miloslav Feistauer

1 General Boundary Value problem

Let us consider the boundary value problem

$$(1) \qquad -\sum_{i=1}^{2} \frac{\partial}{\partial x_i} \left[a_i(x, u(x), \nabla u(x)) \right] + a_0(x, u(x), \nabla u(x)) = f(x) \quad \text{in } \Omega,$$

equipped with the mixed Dirichlet–Neumann boundary conditions

$$(2) \qquad u|_{\Gamma_D} = u_D, \quad \sum_{i=1}^{2} a_i(\cdot, u, \nabla u)\, n_i = \varphi_N \quad \text{on } \Gamma_N.$$

Here $\Omega \subset \mathbb{R}^2$ is a bounded domain with a Lipschitz-continuous boundary $\partial\Omega$ piecewise of the class C^3 and $\partial\Omega = \overline{\Gamma}_D \cup \overline{\Omega}_N$. The sets Γ_D, $\Gamma_N \subset \partial\Omega$ are formed by a finite number of open arcs (i. e., arcs without their endpoints) and $\Gamma_D \cap \Gamma_N = \emptyset$. $\mathbf{n} = (n_1, n_2)$ is unit outer normal to $\partial\Omega$.

Let us introduce the following *basic assumptions on data*:

(A) $\varphi_N \in L^\infty(\Gamma_N)$ is piecewise of the class C^2 on Γ_N, $u_D = u^*|_{\Gamma_D}$, where $u^* \in W^{1,p}(\mathbb{R}^2)$ with $p > 0$, and $\Gamma_D \neq \emptyset$.

Further, we consider a system $\{\Omega_h\}$, $h \in (0, h_0)$, of polygonal approximations of Ω and an open set Ω^* such that $\Omega \cup \Omega_h \subset \Omega^*$ for all $h \in (0, h_0)$ and assume that the following conditions are satisfied: $f \in W^{1,\infty}(\Omega^*)$,

(B) $a_i : \overline{\Omega}^* \times \mathbb{R}^3 \to \mathbb{R}^1$, $i = 0, 1, 2$, are continuous and

$$|a_i(x, \xi)| \leq c_0 \left(1 + \sum_{i=0}^{2} |\xi_i| \right), \quad x \in \overline{\Omega}^*, \ \xi = (\xi_0, \xi_1, \xi_2) \in \mathbb{R}^3.$$

Then it is possible to reformulate problem (1)–(2) in a weak sense. We introduce the forms

$$(3) \qquad a(u, v) = \int_\Omega \left\{ \sum_{i=1}^{2} a_i(\cdot, u, \nabla u) \frac{\partial v}{\partial x_i} + a_0(\cdot, u, \nabla u)\, v \right\} dx,$$

$$(4) \qquad L(v) = L^\Omega(v) + L^\Gamma(v) \equiv \int_\Omega fv\, dx + \int_{\Gamma_N} \varphi_N v\, dS, \quad u, v \in W^{1,2}(\Omega),$$

34

and the space
$$(5) \qquad V = \left\{ v \in W^{1,2}(\Omega);\ v|_{\Gamma_D} = 0 \right\}.$$

We define the *weak solution* to (1)–(2) as a function $u \in W^{1,2}(\Omega)$ satisfying the conditions
$$(6) \qquad u - u^* \in V, \quad a(u,v) = L(v) \quad \forall\, v \in V.$$

The *finite element discretization* of the problem is carried out as it is usual in practice. The domain Ω is approximated by a polygonal domain Ω_h and by $\mathcal{T}_h$ we denote the triangulation of Ω_h with usual properties (see, e. g. [9], [21]). By $\sigma_h = \{P_1, \ldots, P_{Nh}\}$ we denote the set of all vertices of all triangles $T \in \mathcal{T}_h$ and assume that $\sigma_h \subset \overline{\Omega}$, $\sigma_h \cap \partial\Omega_h \subset \partial\Omega$, $\overline{\Gamma}_D \cap \overline{\Gamma}_N \subset \sigma_h$, and the points of $\partial\Omega$, where the C^3-smoothness of $\partial\Omega$ and C^2-smoothness of φ_N is not satisfied, are elements of σ_h. The sets Γ_D, $\Gamma_N \subset \partial\Omega_h$ are approximated in a natural way by Γ_{Dh}, $\Gamma_{Nh} \subset \partial\Omega_h$. We set $h = \max\{\mathrm{diam}(T);\, T \in \mathcal{T}_h\}$.

An approximate solution is sought in the space
$$(7) \qquad X_h = \left\{ v_h \in C(\overline{\Omega}_h);\ v_h|_T \text{ is linear } \forall\, T \in \mathcal{T}_h \right\}.$$

Further, we put
$$(8) \qquad V_h = \left\{ v_h \in X_h;\ v_h|_{\Gamma_{Dh}} = 0 \right\},$$

and approximate the forms a, L by $\tilde{a}_h$, $\tilde{L}_h$ obtained from (3), (4) replacing Ω by Ω_h, Γ_N by Γ_{Nh} and φ_N by its suitable approximation $\varphi_{Nh} : \Gamma_{Nh} \to \mathbb{R}^1$. See [21]. We can put $u_h^* = r_h u^* =$ the Lagrange interpolation of u^*.

The integrals are usually evaluated by means of *numerical integration*. We write
$$(9) \quad \mathrm{a)} \quad \int_{\Omega_h} = \sum_{T \in \mathcal{T}_h} \int_T, \quad \mathrm{b)} \quad \int_T F\,\mathrm{d}x \approx \mathrm{meas}(T) \sum_{j=1}^{k_T} \omega_{T,j} F(x_{T,j})$$

with $\omega_{T,j} \in \mathbb{R}^1$ and $x_{T,j} \in T$. (By $\mathrm{meas}(T)$ we denote the 2-dimensional Lebesgue measure of $T \in \mathcal{T}_h$.) We assume that
$$(10) \qquad \mathrm{a)} \quad \omega_{T,j} > 0, \qquad \mathrm{b)} \quad \sum_{j=1}^{k_T} \omega_{T,j} = 1.$$

The condition b) implies that the quadrature formula (9,b) is of order ≥ 1. (Cf. (17).) Similarly we approximate the integral along Γ_{Nh}.

Applying the numerical integration to the forms $\tilde{a}_h$, $\tilde{L}_h$, we get their approximations a_h, L_h. and introduce the *discrete problem* used in practice: Find $u_h \in X_h$ such that
$$(11) \qquad u_h - u_h^* \in V_h, \quad a_h(u_h, v_h) = L_h(v_h) \quad \forall\, v_h \in V_h.$$

The proof of the existence of an approximate solution is usually easy. It is enough to use the assumption (B) and the *coercivity condition*

$$(D_1) \quad \sum_{i=0}^{2} a_i(x,\xi)\,\xi_i \geq c_1(\xi_1^2 + \xi_2^2) - \left(1 + \sum_{i=0}^{2} |\xi_i|\right), \quad x \in \overline{\Omega}^*, \ \xi \in \mathbb{R}^3,$$

with $c_1 > 0$, $c \geq 0$ independent of x and ξ. See [22] or [20].

In order to establish the *convergence* of approximate solutions to an exact one as $h \to 0$, we consider a *regular* system $\{\mathcal{T}_h\}$, $h \in (0, h_0)$, of triangulations of the domains Ω_h. Main difficulties we meet in the investigation of the convergence are caused by the *nonlinearity* of equation (1) combined with *variational crimes* (see [29]) which we have committed since the domain Ω was approximated by a polygonal domain and numerical integration was used. Moreover, very often we have a lack of monotonicity of the problem and have not sufficient regularity of the exact solution. Therefore, we will discuss the convergence of approximate solutions to an exact one without any assumptions on its regularity. We use the following *main tools* in the study of convergence.

a) *The treatment of the approximation of the curved boundary* $\partial\Omega$ *is* carried out with the aid of the approach worked out by M. Feistauer and A. Ženíšek based on the Zlámal's concept of *ideal interpolation* ([31]). This means that each $v_h \in V_h$ is associated with its *modification* $\hat{v}_h \in V_h$ such that

$$(12) \qquad \|\overline{v}_h - \hat{v}_h\|_{W^{1,2}(\Omega)} \leq c\,h\|v_h\|_{W^{1,2}(\Omega_h)}, \quad h \in (0, h_0).$$

Here $\overline{v}_h$ denotes the natural extension of v_h from $\overline{\Omega}_h$ onto $\overline{\Omega}_h \cup \overline{\Omega}$. Using the representation of u_h in the form $u_h = u_h^* + z_h$, $z_h \in V_h$, we define the modification of u_h:

$$(13) \qquad u_h' = \overline{u}_h^* + \hat{z}_h \in W^{1,2}(\Omega).$$

Furthermore, we use some *auxiliary estimates* over the set $\theta_h = (\Omega - \overline{\Omega}_h) \cup (\Omega_h - \overline{\Omega})$:

$$(14) \ \operatorname{meas}(\theta_h) \leq c\,h^2; \ \|\overline{v}_h\|_{W^{1,p}(\theta_h)} \leq c\,h^{1/p}\|v_h\|_{W^{1,p}(\Omega_h)},$$

$$\|\overline{v}_h\|_{L^p(\theta_h)} \leq c\,h^{2/p}\|v_h\|_{W^{1,p}(\Omega_h)}, \ v_h \in X_h, \ h \in (0, h_0), \ p \in [1, \infty] \ (c = c(p))$$

and moreover the uniform discrete Friedrichs inequality and theorem on traces:

$$(15) \ \|v_h\|_{W^{1,2}(\Omega_h)} \leq c|v_h|_{W^{1,2}(\Omega_h)} = c\left(\int_{\Omega_h} |\nabla v_h|^2 \, dx\right)^{1/2}, \ v_h \in V_h, \ h \in (0, h_0),$$

$$(16) \ \|v_h\|_{L^p(\partial\Omega_h)} \leq c(p)\|v\|_{W^{1,2}(\Omega_h)}, \quad v \in W^{1,2}(\Omega_h), \ h \in (0, h_0).$$

b) *The treatment of the effect of numerical integration* is based on the results by [10] represented by the well-known Theorem 4.1.5 from [9]: *If the quadrature formula* (9,b) *is precise for all polynomials of degree* $\leq 2k - 2$, *then*

$$(17) \qquad \left| \int_T f p \, dx - \operatorname{meas}(T) \sum_{j=1}^{k_T} \omega_{T,j} f(x_{T,j})\, p(x_{T,j}) \right| \leq$$

$$c\, h_T^k (\mathrm{meas}(T))^{1/2-1/q} \|f\|_{W^{k,q}(T)} \|p\|_{W^{1,2}(T)},$$

$$T \in \mathcal{T}_h,\ f \in W^{k,q}(T),\ p = polynomial\ of\ degree\ \leq 1,$$

$$q \in (2, \infty]\ (h_T = \mathrm{diam}(T)).$$

The estimate (17) is applied elementwise with $f := a_i(\cdot, u_h, \nabla u_h)|T$ and either $p := \frac{\partial v_h}{\partial x_i}|T$ or $p := v_h|T$ for $T \in \mathcal{T}_h$, $u_h, v_h \in X_h$. It is clear that the success depends on the behaviour of a_i, namely on its growth and the growth of its derivatives with respect to ξ_0, ξ_1, ξ_2. This fact was first established in [13] and [14], where the following results were proved:

Let us assume that beside (A), (B), (D_1) the following assumptions are satisfied:

(C) $\partial a_i / \partial \xi_j$, $i, j = 0, 1, 2$, are continuous and bounded,

(D_2) $\sum\limits_{i,j=1}^{2} \frac{\partial a_i}{\partial \xi_j}(x, \xi)\, \eta_i \eta_j \geq \alpha(\eta_1^2 + \eta_2^2)$ for $x \in \overline{\Omega}^*$, $\xi \in \mathbb{R}^3$, $\eta = (\eta_1, \eta_2) \in \mathbb{R}^2$ with $\alpha > 0$ independent of x, ξ, η (the condition of pseudomonotonicity),

(E) $\partial a_i / \partial x_k$, $i = 0, 1, 2$, $k = 1, 2$, are continuous and

$$\left| \frac{\partial a_i}{\partial x_k}(x, \xi) \right| \leq c_0 \left(1 + \sum_{j=0}^{2} |\xi_j| \right), \quad x \in \overline{\Omega}^*,\ \xi \in \mathbb{R}^3.$$

Then the error of numerical integration in the form $\tilde{a}_h$ has the estimate

$$(18) \qquad |\tilde{a}_h(u_h, v_h) - a_h(u_h, v_h)| \leq$$
$$\leq\ c\, h \left(1 + \|u_h\|_{W^{1,2}(\Omega_h)} \right) \|v_h\|_{W^{1,2}(\Omega_h)}, \quad u_h, v_h \in X_h,\ h \in (0, h_0).$$

(It easily follows from (17) that the error of numerical integration in the form $\tilde{L}_h$ is of order $O(h)$.)

Further, we have $\overline{u}_h^* \to u^*$ in $W^{1,2}(\Omega)$ and the approximate solutions are uniformly bounded:

$$(19) \qquad \|u_h\|_{W^{1,2}(\Omega_h)} + \|u_h'\|_{W^{1,2}(\Omega)} \leq c \quad \forall h \in (0, h_0).$$

Hence, there exist a sequence $h_n \in (0, h_0)$, $h_n \to 0$ and a function u such that
$$u_{h_n}' \to u \text{ weakly in } W^{1,2}(\Omega) \text{ as } n \to \infty.$$

Using the theory of pseudomonotone operators, we find that $u_{h_n}' \to u$ strongly in $W^{1,2}(\Omega)$, u is the solution of the continuous problem (7) and

$$\|u_{h_n} - \tilde{u}\|_{W^{1,2}(\Omega_{h_n})} \to 0 \quad \text{as } n \to \infty,$$

where $\tilde{u} \in W^{1,2}(\Omega^*)$ is the standard extension of u onto Ω^*.

Writing in (D_2) the sum over i, j from 0 to 2 we obtain the *strongly monotone case*, when the *uniqueness* of the solution holds, and we get the convergence of the whole family $\{u_h\}$, $h \in (0, h_0)$, to the exact solution u. Moreover, we can derive the *error estimate*: Provided $u \in W^{1+\varepsilon,2}(\Omega^*)$ with some $\varepsilon \in (0, 1]$, then the error is of order $O(h^\varepsilon)$. See [21], [23].

All these results were generalized to the equations with discontinuous coefficients in [20] and [30].

The case of a strongly monotone equation with coefficients having general p-growth $(p \geq 1)$ was investigated in [8], but without variational crimes.

2 Stationary Heat Conductivity Problem

Unfortunately, the presented theory does not cover all important situations. This is, e.g., the case of the stationary heat conductivity problem

$$(20) \qquad -\sum_{i=1}^{2} \frac{\partial}{\partial x_i}\left[b(x, u(x))\frac{\partial u(x)}{\partial x_i}\right] = f(x) \quad \text{in } \Omega, \quad u|\partial\Omega = u_D,$$

where the function $b = b(x, \xi_0)$, $x \in \overline{\Omega}^*$, $\xi_0 \in \mathbb{R}^1$, has continuous and bounded derivatives $\partial b/\partial x_k$, $k = 1, 2$, $\partial b/\partial \xi_0$ and there exist constants b_1, b_2 such that

$$(21) \qquad\qquad 0 < b_1 \leq b \leq b_2 \quad \text{in } \overline{\Omega}^* \times \mathbb{R}^1.$$

(For simplicity we consider only the Dirichlet condition.)

Numerical approximations of this problem have been studied by several authors as, e.g., [11], [12], [24], [25] and recently [27], where the uniqueness of the exact solution has also been established. However, the finite element variational crimes are not considered in these papers.

Writing (20) in the form (1), we can see that condition (C) is not satisfied. This causes additional difficulties in estimating the errors caused by variational crimes. Now the effect of numerical integration in the form $\tilde{a}_h$ (yielding the form a_h) is estimated with the use of (17) in a more complicated way than in (18):

$$(22) \quad |\tilde{a}_h(u_h, v_h) - a_h(u_h, v_h)| \leq$$
$$\leq \; c\,h\|u_h\|_{W^{1,2}(\Omega_h)}\|v_h\|_{W^{1,2}(\Omega_h)} + c\,h^{1-2/p}\|u_h\|_{W^{1,2}(\Omega_h)}\|u_h\|_{W^{1,p}(\Omega_h)}\|v_h\|_{W^{1,2}(\Omega_h)},$$
$$u_h, \, v_h \in X_h.$$

Let us note that $p > 2$ and hence, the error tends to zero as $h \to 0$, provided we consider *bounded* families

$$(23) \quad \left\{\|u_h\|_{W^{1,p}(\Omega_h)}\right\}, \quad \left\{\|u_h\|_{W^{1,2}(\Omega_h)}\right\}, \quad \left\{\|v_h\|_{W^{1,2}(\Omega_h)}\right\}, \quad h \in (0, h_0).$$

38

However, we have the boundedness of approximate solutions u_h in $W^{1,2}(\Omega_h)$ only (cf. (19)).

This obstacle was overcome in [17], where we started from the relation $u = u^* + z$, $z \in V$, approximated z by a sequence $z^m \in C^\infty(\overline{\Omega}) \cap V$ (= the space dense in V) and applied (22) with $r_h(u^* + z^m)$ substituted for. Using general approximation results from [10] and estimates (14), we obtain the strong convergence $u_h' \to u$ in $W^{1,2}(\Omega)$ as $h \to 0$.

3 Transonic Potential Flow

In the following part we want to point out that the analysis of the approximation of a curved boundary is not only important for the prove of convergence, but also allows us to obtain a deeper insight into the application of some sophisticated methods.

Transonic potential plane flow is described by the nonlinear mixed type elliptic-hyperbolic the equation

$$(24) \qquad \sum_{i=1}^{2} \frac{\partial}{\partial x_i} \left(\rho(|\nabla u|^2) \frac{\partial u}{\partial x_i} \right) = 0 \quad \text{in } \Omega$$

equipped with the Neumann condition

$$(25) \qquad \rho(|\nabla u|^2) \frac{\partial u}{\partial n} \,|\, \partial\Omega = \varphi_N,$$

where $\rho : [0, \infty) \to [\alpha_0, \alpha_1]$, α_0, $\alpha_1 > 0$ has a special concrete form. The weak solution is defined by (6) with $V = \{v \in W^{1,2}(\Omega); \int_\Omega v \, dx = 0\}$, $u^* = 0$ and the form a corresponding to the equation (24). Similarly as in (11) we write down the discrete problem. Unfortunately, there is no monotonicity or compactness in the continuous problem and therefore, its solvability still resists the effort of many mathematicians.

Since transonic flow is an irreversible process, it is necessary to include the second law of thermodynamics represented by the entropy inequality and the condition of bounded velocity. The discrete forms of these conditions read

$$(26) \qquad -\int_{\Omega_h} \nabla u_h \cdot \nabla v_h \, dx - K \int_{\Omega_h} v_h \, dx \le 0$$

$$\forall v_h \in X_{h0}^+ := \{v_h \in X_h; v_h \ge 0 \text{ on } \Omega_h \text{ and } v_h = 0 \text{ on } \partial\Omega_h\},$$

$$(27) \qquad |\nabla u_h|^2 \le s^* \quad \text{in } \Omega_h,$$

$K = \text{const..}$ For a detailed explanation, see [14, Chap. 6].

The difficulties connected with finding an approximate solution satisfying (26), lead us to the use of least squares with penalization as was proposed in

[1] (cf. also [26]) and further developed in [6]. Let us denote by w_{hi} the basis functions in X_h such that $w_{ih}(P_j) = \delta_{ij}$. Then we set $A_i = \operatorname{meas}(\operatorname{supp} w_{hi})$ and define the functional $J_h : X_h \to \mathbb{R}^1$:

$$(28)\quad J_h(\varphi_h) = \frac{1}{2} \int_{\Omega_h} |\nabla \xi_h(\varphi_h)|^2 \, \mathrm{d}x +$$

$$+ \frac{\mu}{2} \sum_{P_i \in \sigma_h \cap \Omega_h} \frac{1}{A_i^\varepsilon} \left[\left(-\int_{\Omega_h} \nabla \varphi_h \cdot \nabla w_{hi} \, \mathrm{d}x - K \int_{\Omega_h} w_{hi} \, \mathrm{d}x \right)^+ \right]^2 .$$

Here $\mu, \varepsilon > 0$, $K \geq 0$ are fixed constants. The function $\xi_h = \xi_h(\varphi_h)$ is the solution of the discrete problem

$$(29)\qquad \xi_h \in V_h,$$

$$\int_{\Omega_h} \nabla \xi_h \cdot \nabla v_h \, \mathrm{d}x = a_h(u_h, v_h) - L_h(v_h) \quad \forall v_h \in V_h.$$

Now we define an *approximate physical solution* as a function $u_h \in V_h$ satisfying (27) and

$$(30)\qquad u_h = \operatorname*{arg\,min}_{\varphi_h \in V_h, |\nabla \varphi_h|^2 \leq s^*} J_h(\varphi_h).$$

The convergence theory is the subject of several papers. It uses the entropy compacfitication of the transonic flow problem investigated in [2], [3], [4], [5], [7], [18], [19], [28]. In particular, in [2], [3] the convergence $u_h \to u$ (= the exact physically admissible solution whose existence and uniqueness has to be, of course, assumed) was proved under the assumptions that the domain Ω is polygonal and $\varepsilon > 1$.

The more realistic case of curved boundary $\partial \Omega$ approximated by $\partial \Omega_h$ was treated in [4], [5]. It is interesting that the analysis of the convergence and, particularly, the investigation of the effect of the approximation of the curved boundary leads to the identification of the parameter ε. Namely, the convergence proof based on the generalization of the approach from [21], [22] requires the condition $\varepsilon \in [1, 3/2)$. Recently this bound has been made less sharp, i.e., $\varepsilon \in [1, 2)$ in [5] which uses similar, but slightly improved technique. By H. Berger, the theoretical results are in agreement with numerical experiments that yield the best results for $\varepsilon \in (1.1., 1.3)$.

4 Boundary Value Problem with a Nonlinear Newton Condition

In the end we will be concerned with the effect of numerical integration in nonlinear Newton boundary condition arising in modelling of an electrolysis

40

process. Let us consider here a simplified model problem

$$(31) \qquad \Delta u = \operatorname{div} \mathbf{f} \quad \text{in } \Omega, \quad \frac{\partial u}{\partial n} + k|u|^\alpha u = f_n = \mathbf{f} \cdot \mathbf{n} \quad \text{on } \partial\Omega,$$

with given constants $k > 0$, $\alpha \geq 0$ and a given vector function $\mathbf{f} : \overline{\Omega}^* \to \mathbb{R}^2$. $\partial/\partial n$ denotes the derivative in the direction of unit outer normal $\mathbf{n}$ to $\partial\Omega$.

In [16] the unique solvability and the convergence of approximate finite element solutions was established, provided the domain Ω is polygonal and the integrals in the nonlinear boundary form are evaluated exactly. We meet difficulties in the analysis of numerical integration in this nonlinear boundary form

$$\tilde{c}_h(u_h, v_h) = k \sum_{S \subset \partial\Omega_h} \int_S |u_h|^\alpha u_h \, v_h \, dS,$$

where $S \subset \partial\Omega_h$ are sides of triangles $T \in \mathcal{T}_h$ adjacent to $\partial\Omega_h$. From (17) it is obvious that the crucial role is played by a suitable estimate of the term $h_S^k \|\,|u_h|^\alpha u_h\|_{W^{k,q}(S)}$. Let us consider, e.g., $\alpha > 1/2$ and $q = 2$. Then, using $k = 1$, the inequality $\|v_h\|_{1,\partial\Omega_h} \leq c\,h^{-1/2}\|v_h\|_{1,\Omega_h}$, $v_h \in X_h$, the inverse inequality and the uniform boundedness of u_h in $W^{1,2}(\Omega_h)$, we find that the error of numerical integration is of order $O(1)$. Hence, the convergence is not guaranteed.

Therefore, we try to "overintegrate", i.e., we use a quadrature formula with $k = 2$. Then we can find that the numerical integration error is of order $O(h^{1-\alpha/r})$ with $r \geq 1$, which yields the *convergence for any* $r \in (\alpha, \infty)$. We see that we "lose more than one order" because of the nonlinearity on the boundary. The detailed analysis will be the subject of [15].

As the conclusion let us note that all the cited results were obtained for piecewise *linear* conforming finite elements only. The *extension to higher degree finite elements remains open*.

References

[1] Bristeau, M.O., Glowinski, R., Periaux, J., Perrier, P., Pironneau, P., Poirier, G.: Application of optimal control and finite element methods to the calculation of transonic flows and incompressible viscous flows. In: Numerical Methods in Applied Fluid Dynamics (B. Hunt, ed.). Academic Press, London, 1980, 203–312.

[2] Berger, H.: Finite-Element-Approximationen für transonische Strömungen. Dr. rer. nat. Dissertation, Universität Stuttgart, 1989.

[3] Berger, H.: A convergent finite element formulation for transonic flow. Numer. Math. **56**, 425–447, 1989.

[4] Berger, H., Feistauer, M.: Analysis of the finite element variational crimes in the numerical approximation of transonic flow. Bericht Nr. 27, Nov. 90, Seminar

Analysis und Anwendungen, Universität Stuttgart (to appear in Math. Comp. 1993).

[5] Berger, H., Feistauer, M.: Mathematical theory of the finite element solution of transonic flow. In: Methoden und Verfahren der mathematischen Physik **37**: Direct and Inverse Boundary Value Problems (R. Kleinman, R. Kress, E. Martensen, eds.), P. Lang, Frankfurt am Main – Bern – New York – Paris, 1991, 15–25.

[6] Berger, H., Warnecke, G., Wendland, W.: Finite elements for transonic flows. Numer. Methods for P.D.E **6**, (1990), 17–42.

[7] Berger, H., Warnecke, G., Wendland, W.: Analysis of a FEM/BEM Coupling Method for Transonic Flow Computations. Preprint, Universität Stuttgart, 1993.

[8] Chow, S. S.: Finite element error estimates for non-linear elliptic equations of monotone type. Numer. Math. **54**, (1989), 373–393.

[9] Ciarlet, P. G.: The Finite Element Method for Elliptic Problems. North–Holland, Amsterdam, 1979.

[10] Ciarlet, P. G., Raviart, P. A.: The combined effect of curved boundaries and numerical integration in isoparametric finite element method. In: The Mathematical Foundations of the Finite Element Method with Applications to Partial Differential Equations (A. K. Aziz, ed.), Academic Press, New York, 1972, 409–474.

[11] Douglas, J.: H^1-Galerkin methods for a nonlinear Dirichlet problem. In: Proc. Conf. on Mathematical Aspects of Finite Element Methods, Lecture Notes in Math., Vol. **606**, Springer–Verlag, Berlin 1977, 64–86.

[12] Douglas, J., Dupont, T.: A Galerkin method for a nonlinear Dirichlet problem. Math. Comp. **29**, (1975), 689–696.

[13] Feistauer, M.: On the finite element approximation of a cascade flow problem. Numer. Math. **50**, (1987), 655–684.

[14] Feistauer, M.: Mathematical Methods in Fluid Dynamics. Pitman Monographs and Surveys in Pure and Applied Mathematics 67, Longman Scientific & Technical, Harlow, 1993.

[15] Feistauer, M.: Finite element approximation of a problem with nonlinear Newton condition (in preparation).

[16] Feistauer, M., Kalis, H., Rokyta, M.: Mathematical modelling of an electrolysis process. Comment. Math. Univ. Carolinae **30**, (1989), 465–477.

[17] Feistauer, M., Křížek, M., Sobotíková, V.: An analysis of finite element variational crimes for a nonlinear elliptic problem of a nonmonotone type. Submitted to East-West J. Numer. Anal.

[18] Feistauer, M., Mandel, J., Nečas, J.: Entropy regularization of the transonic potential flow problems. Comment. Math. Univ. Carolinae **25**, (1984), 431–443.

[19] Feistauer, M., Nečas, J.: On the solvability of transonic potential flow problems. Z. Anal. Anw. 4, (1985), 305–329.

[20] Feistauer, M., Sobotíková, V.: Finite element approximation of nonlinear elliptic problems with discontinuous coefficients. RAIRO Modél. Math. Anal. Numer. (M^{2}AN) **24**, (1990), 457–500.

[21] Feistauer, M., Ženíšek, A.: Finite element solution of nonlinear elliptic problems. Numer. Math. **50**, (1987), 451–475.

[22] Feistauer, M., Ženíšek, A.: Compactness method in the finite element theory of nonlinear elliptic problems. Numer. Math. **52**, (1988), 147–163.

[23] Feistauer, M., Ženíšek, A.: Finite element variational crimes in nonlinear elliptic problems. In: Proc. of the ISNA 87, Teubner Texte zur Mathematik, Band 107, Lepzig, 1988, 28–35.

[24] Frehse, J., Rannacher, R.: Optimal uniform convergence for the finite element approximation of a quasilinear elliptic boundary value problem. In: Proc. of the U.S.–German Symposium "Formulations and Computational Algorithms in Finite Element Analysis", M.I.T, 1976.

[25] Frehse, J., Rannacher, R.: Asymptotic L^∞-error estimates for linear finite element approximations of quasilinear boundary value problems. SIAM J. Numer. Anal. **15**, (1978), 418–431.

[26] Glowinski, R.: Numerical Methods for Nonlinear Variational Problems. Springer–Verlag, New York – Berlin – Heidelberg – Tokyo, 1984.

[27] Hlaváček, I., Křížek, M., Malý, J.: On Galerkin approximations of a quasilinear nonpotential elliptic problem of a nonmonotone type. J. Math. Anal. Appl. (to appear).

[28] Mandel, J., Nečas, J.: Convergence of finite elements for transonic potential flow. SIAM J. Numer. Anal. **24**, (1987), 985–996.

[29] Strang, G.: Variational crimes in the finite element method. In: The Mathematical Foundations of the Finite Element Method with Applications to Partial Differential Equations (A.K. Aziz, ed.), Academic Press, New York, 1972, 689–710.

[30] Ženíšek, A.: The finite element method for nonlinear elliptic equations with discontinuous coefficients. Numer. Math. **58**, (1990), 51–77.

[31] Zlámal, M.: Curved elements in the finite element method I. SIAM J. Numer. Anal. **10**, (1973), 229–240.

This work was partially supported by the grant 201/93/2177 of the Czech Grant Agency.

Miloslav Feistauer
Faculty of Mathematics and Physics
Charles University Prague
Sokolovská 83, 186 00 Praha 8, Czech Republic
e-mail: Feist@cspguk11.bitnet

Elasticity Boundary Value Problems on Cracks with Partial Contact of Their Surfaces

Robert Goldstein, Yuriĭ Zhitnikov

Abstract

A class of three-dimensional elasticity problems of equilibrium of cavities and crack-cuts in a homogeneous, isotropic, linear elastic space is considered, assuming that their surfaces may come into contact under the action of a bulk force system. The contact region boundaries are *a priorily* unknown. In the x_1, x_2, x_3 coordinate system the cross section with the plane $x_3 = 0$ occupies a domain Ω and the distance between the surfaces of the cavity $W_o(x_1, x_2)$ is a single-valued function of (x_1, x_2). It is small compared to the size of Ω (narrow cavity).

The original problem is divided into a symmetric and antisymmetric one relative to the plane $x_3 = 0$. In the first case there is a discontinuity of the normal displacement along the Ω plane, while in the second case there is a discontinuity of the tangential displacement [1]. Such a representation is related to the fact that the normal and shear components of the displacement discontinuity do not produce shear and normal displacements in the plane $x_3 = 0$ of the corresponding infinite elastic space. The present paper deals with the case of bulk forces symmetric to the plane $x_3 = 0$. The problem to determine the normal discontinuity of the displacement and the boundary between contact and opening regions is called the normal problem.
A series of statements on the properties of the solution to the normal problem and its uniqueness are proved. The analysis of the properties of the solution to the normal problem is based on the positivity of the solution of the normal problem [2]. Incorporating the obtained properties of the normal problem, a new regular numerical-analytical method for the construction of the solution with contact regions is suggested.
The approaches used to build a solution for the problem of a cavity with contact regions included the variational method [I-4] and the method of construction of a nonsingular solution [5-7]. The present work demonstrates the equivalence of the two approaches with the approach proposed.

1 Formulation of the normal problem and uniqueness of the solution

Let us consider the equilibrium of a linearly elastic space containing a cavity (in particular a crack-cut). Assume that the cross section of the cavity with the plane $x_3 = 0$ is a domain Ω. Let the distance between the surfaces of the cavity $W_o(x_1, x_2)$ be a single-valued function of $(x_1, x_2) \in \Omega$ which is small compared with the characteristic size of Ω (a narrow cavity).

We will bring the conditions at the surfaces of the cavity onto the plane $x_3 = 0$. Assume that the contact region is formed in the Ω plane under the action of a bulk force system, these forces being symmetric to the plane $x_3 = 0$. The boundary conditions at the plane $x_3 = 0$ have the form

$$\sigma_{33}^{\pm} = 0, \ (x_1, x_2) \in \Omega \setminus F, \ \sigma_{33}^{\pm} \leq 0, \ (x_1, x_2) \in F$$

$$u_3^+ - u_3^- = -W_o(x_1, x_2), \ (x_1, x_2) \in F \tag{A}$$

$$u_3^+ - u_3^- = 0, \ (x_1, x_2) \in R^2 \setminus \Omega, \ x_3 \to \pm 0$$

$$u_3(x_1, x_2) \geq -W_o(x_1, x_2), \ (x_1, x_2) \in \Omega.$$

Here, F is the region where the cavity surfaces are in contact (hence F is the contact region). Outside the domain Ω, the external loads are given in the form of a bulk force distribution with density $\rho(x_1, x_2, x_3)$. If $W_o(x_1, x_2, x_3) \equiv 0$, the cavity becomes a crack-cut.

When solving the boundary-value problem (A), it is convenient to replace the system of external loads by boundary conditions at the plane $x_3 = 0$. To do this, we will find the distribution of stresses in a solid in the domain $\Omega : -\sigma_{33}^o(x_1, x_2)$, assuming that the corresponding elasticity problem for a solid without a cavity is already solved. Then we add the stresses $-\sigma_{33}^o(x_1, x_2)$ with opposite sign to the boundary conditions for stresses in the domain Ω [2]. When this procedure is completed, the boundary conditions (A) become

$$W(x_1, x_2) = u_3^+(x_1, x_2) - u_3^-(x_1, x_2) = -W_o(x_1, x_2), \ (x_1, x_2) \in F \tag{1.1}$$

$$W(x_1, x_2) \geq -W_o(x_1, x_2), \ (x_1, x_2) \in \Omega \tag{1.2}$$

$$\sigma_{33}(x_1, x_2) \leq \sigma_{33}^o(x_1, x_2), \ (x_1, x_2) \in F \tag{1.3}$$

$$\sigma_{33}(x_1, x_2) = \sigma_{33}^o(x_1, x_2), \ (x_1, x_2) \in \Omega \setminus F. \tag{1.4}$$

The contact region F (correspondingly, opening region $D = \Omega \setminus F$), should be determined through the process of solving the problem. It should be noted that, due to the symmetry of the problem, its solution satisfies the conditions $u_3^+(x_1, x_2) = u_3^-(x_1, x_2)$, $(x_1, x_2) \in R^2 \setminus \Omega$, $\sigma_{3i}^\pm(x_1, x_2) = 0$ for $x_3 = 0 (i = 1, 2)$. Let Γ denote the boundary of the domain D.

We now consider the question of uniqueness of the solution to the boundary-value problem (1.1)-(1.4). Assume that the solution of the mixed problem of elasticity for a half-space with line Γ separating the boundary conditions is unique [8]. In this case, the solution of the problem for a cavity with contact and opening regions and acting system of bulk forces is unique.

Proof. Let us suppose that two opening regions D_1 and D_2 exist with discontinuities of displacements $W^{(1)}, W^{(2)}$, stress states $\sigma_{ik}^{(1)}, \sigma_{ik}^{(2)}$ and with displacements $u_i^{(1)}, u_i^{(2)}$ at the action of the given system of bulk forces. The assumed uniqueness of the solution of the mixed elasticity problem with separating line Γ implies that the regions D_1 and D_2 cannot coincide. Let their intersection be $D_o : D_1 \cap D_2$. It follows from the conditions in the contact region (1.1) that

$$\sigma_{33}^{(1)} \leq \sigma_{33}^o, \ (x_1, x_2) \in D_2 \setminus D_o; \ \sigma_{33}^{(2)} \leq \sigma_{33}^o, \ (x_1, x_2) \in D_1 \setminus D_o. \tag{1.5}$$

We subtract the parts corresponding to the first and second states. Taking into account (1.1)-(1.4), (1.5), we obtain in the plane $x_3 = 0$

$$\vartheta_3 \equiv \vartheta_3^+ - \vartheta_3^- = W^{(1)} + W_o \geq 0, \ \Sigma_{33} = \sigma_{33}^o - \sigma_{33}^{(2)} \geq 0, \ (x_1, x_2) \in D_1 \setminus D_o$$

$$\vartheta_3 = W^{(1)} - W^{(2)}, \ \Sigma_{33} = 0, \ (x_1, x_2) \in D_o \tag{1.6}$$

$$\vartheta_3 = -W^{(2)} - W_o \leq 0, \ \Sigma_{33} = \sigma_{33}^{(1)} - \sigma_{33}^o \leq 0, \ (x_1, x_2) \in D_2 \setminus D_o$$

$$\Sigma_{ik} = \sigma_{ik}^{(1)} - \sigma_{ik}^{(2)}, \ \vartheta_i = u_i^{(1)} - u_i^{(2)}.$$

The elastic energy for the stress state Σ_{ik} with given stresses and displacements at the boundary (1.6) is nonpositive

$$W = \frac{1}{2} \int_\Omega \Sigma_{33} \vartheta_3 n_3^+ \, dx_1 \, dx_2 \leq 0, n_3^+ = -1,$$

this is impossible [8]. Thus, $W_3^{(1)} = W_3^{(2)}$ and the uniqueness of the solution is proved.

A similar statement holds for a crack-cut $(W_o(x_1, x_2) \equiv 0, (x_1, x_2) \in \Omega)$ with contact and opening-regions.

When studying the properties of the solution to the normal problem, we will use the positivity of the solution [2]: The discontinuity of the displacement is nonnegative, $W(x_1, x_2) \geq 0$, for any point (x_1, x_2) in a domain Ω of the plane $x_3 = 0$, provided that $\sigma_{33}^\pm(x_1, x_2) \leq 0$, $(x_1, x_2) \in \Omega$.

We now use positivity [2] and uniqueness of the solution to prove some of its properties.

2 Properties of the solution to the normal problem

Together with the original problem (1.1)-(1.4), we will consider the mixed problem (1.1),(1.2),(1.4) i.e. the problem where the constraint on the stress σ_{33} are omitted in the region $\Omega \setminus D$. The solution of this boundary-value problem is not unique in general, since the constraint (1.3) is removed. For given loads, to each solution of the problem (1.1),(1.2),(1.4) there is a corresponding opening region $D_k^{(i)}$ for the cavity surfaces (where (1.2) holds) and a contact region $F_k^{(i)}$ (where (1.3) may not hold). Let K be a set of solutions of the problem (1.1),(1.2),(1.4). It contains, in particular, the solution of the original problem (1.1)-(1.4).

For equilibrium states from the set K, the following statement holds due to the uniqueness theorem for the solution of the problem (1.1)-(1.4).

Corollary 1. Consider the solution of the problem (1.1)-(1.4) for a cavity with cross section Ω (in the plane $x_3 = 0$), being its contact region F and its opening region D, when given loads are applied. Let K be the set of equilibrium states introduced above. Then, for any element from K, subdomains $F_K' \subset F_K$ exist in the contact region F_K, where

$$\sigma_{33}^{(K)}(x_1, x_2) > \sigma_{33}^o(x_1, x_2), \ (x_1, x_2) \in F_K',$$

if the opening region for D_K surfaces does not coincide with D. Let us suppose the contrary, i.e. $\sigma_{33}^{(K)} \leq \sigma_{33}^o, \ \forall (x_1, x_2) \in F_K$. This means that the constraint (1.3) is fulfilled in the region F_K and F_K, D_K are the cavity's contact and opening regions for the problem (1.1)-(1.4), just as are F, D.

Statement 1. Consider an equilibrium state of a cavity with cross section Ω, contact region F, and opening region D together with given applied loads. Then the opening region D_K, corresponding to any arbitrary equilibrium state from the set K and not coinciding with D, is contained in $D, D_K \subset D$.

Proof. Suppose that there is an equilibrium state from the set K for which the opening region $D_K = D_K^{(1)}$ is not contained in D. According to Corollary 1, since $D_K^{(1)}$ does not coincide with D, in the contact region $F_K^{(1)} = \Omega \setminus D_K^{(1)}$ there should be subdomains $F_K^{(1)'} \subset F_K^{(1)}$ where $\sigma_{33}^{(1)}(x_1, x_2) > \sigma_{33}^o(x_1, x_2), \ (x_1, x_2) \in F_K^{(1)'}$. Now consider another equilibrium state from the set K (with the same applied loads) with corresponding opening region $D_K^{(2)} = D_K^{(1)} \cup F_K^{(1)}$. We now show that

$$W^{(2)}(x_1, x_2) = W^{(1)}(x_1, x_2) + \delta W^{(1)}(x_1, x_2) \geq W^{(1)}(x_1, x_2), (x_1, x_2) \in D_K^{(2)}$$

holds for the discontinuity of the displacement.

Indeed, we may represent the equilibrium state with opening region $D_K^{(2)}$ and given loads as a sum of two equilibrium states of this cavity. The first one is the equilibrium state with opening region $D_K^{(1)}$ and the same loads. The second one is the equilibrium state with opening region $D_K^{(2)}$ and stresses $\delta\sigma_{33}^{(1)} = \sigma_{33}^o - \sigma_{33}^{(1)}$, acting only at subdomains $F_K^{(1)'}$ and having the same absolute values as the stresses occurring at $F_K^{(1)'}$ in the first state, but with the opposite sign.

Since $\sigma_{33}^{(1)} \geq \sigma_{33}^o$, $\delta\sigma_{33}^{(1)} < 0$, $(x_1, x_2) \in F_K^{(1)'}$, we have $\delta W^{(1)}(x_1, x_2) \geq 0$, $(x_1, x_2) \in D_K^{(2)}$ according to the property of positivity of the solution $\delta W^{(1)}$ corresponding to the second state. Thus, it holds $W^{(2)}(x_1, x_2) \geq W^{(1)}(x_1, x_2)$, $(x_1, x_2) \in D_K^{(2)}$ for equilibrium states from the set K with opening regions $D_K^{(1)}$, $D_K^{(2)}$ and given loads. Since $W^{(1)}(x_1, x_2) = W^{(2)}(x_1, x_2)$, $(x_1, x_2) \in F^{(2)}$ for every point in the entire $\Omega = F_K^{(2)} \cup D_K^{(2)}$, we end up with $W^{(2)}(x_1, x_2) \geq W^{(1)}(x_1, x_2)$.

If the domain D_K is not contained in D (and in particular, does not coincide with it), then we obtain an equilibrium state with opening region $D_K^{(3)} = D_K^{(2)} \cup F_K^{(2)'}$ repeating the procedure used for constructing the equilibrium state with opening region $D_K^{(2)}$. Here $F_K^{(2)'}$ is a subdomain, where $\sigma_{33}^{(2)}(x_1, x_2) \geq \sigma_{33}^o(x_1, x_2)$, corresponding $W^{(3)}(x_1, x_2) \geq W^{(2)}(x_1, x_2)$, $\forall(x_1, x_2) \in \Omega$ and so on.

We finally obtain a sequence of equilibrium states from the set K with opening regions $D_K^{(1)} \subset D_K^{(2)} \subset \cdots \subset D_K^{(n)} \subset \cdots \subset \Omega$ so that $W^{(1)}(x_1, x_2) \leq W^{(2)}(x_1, x_2) \leq \cdots \leq W^{(n)}(x_1, x_2) \leq \cdots$, $(x_1, x_2) \in \Omega$. Since the regions $D_K^{(i)}$ form a sequence of expanding regions and are all contained in Ω, the domain $\bigcup_{i=1}^{\infty} D_K^{(i)} = D' \subset \Omega$ exists. By construction, $\sigma_{33}(x_1, x_2) \leq \sigma_{33}^o(x_1, x_2)$, $\forall(x_1, x_2) \in F' = \Omega \setminus D'$, i.e. the domain F' is the contact region. We have built an equilibrium state of a cavity with opening region D' and contact region F', which satisfies the constraints $W(x_1, x_2) \geq -W_o(x_1, x_2), (x_1, x_2) \in D'$, $\sigma_{33}(x_1, x_2) \leq \sigma_{33}^o(x_1, x_2)$, $(x_1, x_2) \in F'$. Since D' and F' do not coincide with D and F respectively, this means that another solution for the original problem (1.1) (1.4) has been derived, which contradicts the uniqueness theorem.

Consequently, in the set K of equilibrium states there does not exist a state with opening region D_K not contained in D. Thus, if D_K does not coincide with D, it is contained in D. Statement 1 is proved.

48

A similar statement holds for crack-cuts.

Statement 2. If the opening region D_1 corresponds to the crack-cut $\Omega_1 \subset \Omega$ with the same system of acting loads, then $D_1 \subset D$, where D is an opening region of the crack Ω. (Compare with Corollary 3 of Statement 1).

From statement 1 it follows that for any equilibrium state from the set K with opening region $D_K^{(1)}$ not coinciding with D and $D_K^{(1)} \subset D$, according to the algorithm described in the proof, one may build a sequence of equilibrium states with opening regions $D_K^{(i)} (i = 1, 2, \cdots)$ containing each other, for which the discontinuities of displacement in each point satisfy inequalities

$$W^{(1)}(x_1, x_2) \leq \cdots \leq W^{(n)}(x_1, x_2) \leq \cdots , (x_1, x_2) \in \Omega.$$

All regions in the sequence $D_K^{(i)}$ are contained in $D = \bigcup_{i=1}^{\infty} D_K^{(i)}$ and the sequence $W^{(i)}(x_1, x_2)$ is bounded from above.

Since the sequence

$$D_K^{(i)} = \bigcup_{j=1}^{i} D_K^{(j)}$$

tends to

$$D = \bigcup_{j=1}^{\infty} D_K^{(j)}$$

as $i \to \infty$, it follows that

$$\lim_{i \to \infty} W^{(i)}(x_1, x_2) = W(x_1, x_2), \ (x_1, x_2) \in D_K^{(i)} \subset D.$$

Indeed, let $\Delta W^{(i)} = W - W^{(i)}, \Delta \sigma_{33} = \sigma_{33} - \sigma_{33}^{(i)}$. Consider the difference of stressed states corresponding to the discontinuities of the displacements $W^{(i)}$ and W. The boundary conditions corresponding to this difference at the plane R^2 have the form

$$\Delta \sigma_{33}(x_1, x_2) = 0, \ (x_1, x_2) \in D^{(i)}$$

$$\Delta \sigma_{33}(x_1, x_2) = \sigma_{33} - \sigma_{33}^{(i)}, \ (x_1, x_2) \in D \setminus D^{(i)}$$

$$\Delta W^{(i)}(x_1, x_2) = 0, \ (x_1, x_2) \in R^2 \setminus D.$$

Since $(D \setminus D_k^{(i)} \to \leq)$ as $i \to \infty$, the solution of this boundary-value problem $\Delta W^{(i)} \to 0$ for $i \to \infty$. Thus,

$$\lim_{i \to \infty} W^{(i)}(x_1, x_2) = W(x_1, x_2).$$

From this fact and the fact that $\{W^{(i)}\}$, $(x_1, x_2) \in \Omega$ is non-decreasing, it follows that $\forall_i$, $W^{(i)}(x_1, x_2) \leq W(x_1, x_2)$, $(x_1, x_2) \in \Omega$. Moreover, $W^{(i)}(x_1, x_2) = -W^\circ(x_1, x_2)$, $(x_1, x_2) \in F$.

Thus,

$$W^{(1)}(x_1, x_2) \leq W^{(2)}(x_1, x_2) \leq \cdots \leq W^{(i)}(x_1, x_2) \leq \cdots$$
$$\leq W(x_1, x_2), \ (x_1, x_2) \in \Omega.$$

We have proved the following statement.

Statement 3. For a cavity with cross-section Ω (for $x_3 = 0$) and given external loads, the discontinuity of displacement of its surfaces reaches its maximum pointwise in $(x_1, x_2) \in \Omega$ at the solution of problem (1.1)-(1.4).

Remark. It follows from Statement 3 that for a crack in the equilibrium state with surfaces being partially in contact with each other its volume is maximal compared to the volume it would have in the case of different contact regions and the same loads. This property is established in a different way in [9].

Statement 4. Regard a solution of the problem on the cavity Ω with given loads, built without taking into account the constraints $W(x_1, x_2) \geq -W_o(x_1, x_2)$, $(x_1, x_2) \in \Omega$. Let $F_o \subset \Omega$ be the region such that $W(x_1, x_2) \leq -W_o(x_1, x_2)$, $(x_1, x_2) \in F_o$ (the region of "overlapping" of the surfaces). Then F is contained within the limits of F_o for this solution, if these constraints are taken into account and the loads are the same at the contact region.

Proof. Consider three possible cases of the equilibrium of the cavity Ω with acting system of external loads.

1°. The equilibrium of the cavity Ω with given acting external load and without the constraint on the discontinuity of the displacement $W(x_1, x_2) \geq -W_o(x_1, x_2), (x_1, x_2) \in \Omega$. We find that the cavity's surfaces "overlap" in the region F_o (we denote the discontinuity of the displacement in this case $W^\circ(x_1, x_2)$, $W^\circ(x_1, x_2) \leq -W_o(x_1, x_2)$, $(x_1, x_2) \in F_o$).

2°. The equilibrium of the cavity Ω with the constraints on the discontinuity of the displacement $W(x_1, x_2) \geq -W_o(x_1, x_2)$, $(x_1, x_2) \in \Omega$ and on the stress in the contact region.

3°. The equilibrium state of the cavity Ω, when external loads are absent, the constraint on the discontinuity of the displacement is neglected, the stresses acting in the region $F \subset \Omega$ are $\sigma_{33}^o - \sigma_{33} \geq 0$, and the loads are absent in the opening region D:

50

In this case the discontinuity of the displacement satisfies

$$\delta W(x_1, x_2) \leq 0, \ (x_1, x_2) \in \Omega$$

according to the positivity property of the solution [2]. In case 1°, the boundary-value problem may be represented as a sum of the boundary-values 2° and 3°. Then the discontinuity of the displacement in the cavity Ω in case 1° can be represented as the sum: $W^\circ(x_1, x_2) = W(x_1, x_2) + \delta W(x_1, x_2), \ (x_1, x_2) \in \Omega$. Since $W(x_1, x_2) = -W_o(x_1, x_2), (x_1, x_2) \in F$, and $\delta W(x_1, x_2) < 0, (x_1, x_2) \in \Omega \setminus \partial\Omega$, we have $W^\circ(x_1, x_2) = -W_o(x_1, x_2) + \delta W(x_1, x_2) < -W_o(x_1, x_2), \ (x_1, x_2) \in F$, hence mutual penetration of cavity surfaces should occur in the region F. Consequently, $F \subset F_o$, which was to be proved.

3   Determination of unknown boundaries

The method to find unknown boundaries was in fact described in section 1 during the proof of Statements 1 and 2. Indeed, let $F_K^{(1)}$ and $D_K^{(1)}$ be the regions of contact and opening corresponding to some equilibrium state from the set K. Then, as it has been proved, $D_K^{(1)} \subset D$ (Statement 1). Constructing the expanding sequence of regions $D_K^{(i)}$ we obtain in the limit the required opening region for the normal problem D. The region overlapping $F^{(1)}$ may be taken as a first approximation for the contact region, and, correspondingly, $D^{(1)} = \Omega \setminus F^{(1)}$ for the opening region. The contact region F is contained in F_o according to Statement 3. Let's suppose that in the region $D^{(1)} = \Omega \setminus F^{(1)}(F^{(1)} = F_o)$ the condition of the absence of overlaping holds $W(x_1, x_2) \geq -W_o(x_1, x_2)$, $(x_1, x_2) \in D^{(1)}$. Then the equilibrium state with opening region $D^{(1)}$ and contact region $F^{(1)}$ belongs to the set K, and consequently $D^{(1)} \subset D$ holds according to Statement 1. Then, we build the sequence of contact regions $F_K^{(i)}$ (or the sequence of the opening regions $D_K^{(i)}$) that tends to D according to the algorithm from section 1, Statement 1. We take $D_K^{(1)} = D^{(1)}$ and $F_K^{(1)} = F^{(1)}$ as the first step, then choose F_k^i such that $F_k^i \subset F_k^{i+1}$. Since in the contact region $F = \Omega \setminus D$ the constraint $\sigma_{33}(x_1, x_2) \leq \sigma_{33}^o(x_1, x_2), \ (x_1, x_2) \in F$ holds, it can be shown that the boundary Γ between regions D and F is a line where the solution is non-singular (see below). Thus the proposed regular algorithm may be used to search the unknown boundary between the regions of contact and opening with a non-singular solution. A similar method of searching the unknown boundary between contact and opening regions may be suggested for the crack-cut of Ω.

Indeed, let's suppose that the distribution of normal stresses $\sigma_{33}^o(x_1, x_2)(x_1, x_2) \in \Omega$ is known in a solid along the crack plane. We choose a region $D^{(1)}$ where $\sigma_{33}^o(x_1, x_2) > 0$, $(x_1, x_2) \in D^{(1)} \subset \Omega$. It follows from the positivity property of the solution [2] that the discontinuity of the displacement exists and $W(x_1, x_2) \geq 0$, $(x_1, x_2) \in D^{(1)}$. Thus, the equilibrium state with opening region $D^{(1)}$ belongs to the set K. Then we take $D_K^{(1)} = D$ as first approximation for the opening region and build an expanding sequence of opening regions $D_K^{(i)}$ according to the algorithm described in section 1. We finally obtain the opening region for the normal problem.

The difference of the methods used for a cavity and a crack-cut is that in the first case we know the contact region F_K for the equilibrium state from the set K at the first step of the iteration process (the region of overlaping), while in the second case-the opening region from the set K is known.

The described regular method for searching the boundary between the contact and opening regions may be used for solving three-dimensional problems of planar cracks in situations where the positivity property of the solution takes place.

4 Properties of the elastic energy functional

Using the results of section 2 we now consider certain properties of the elastic energy functional in the problem of a crack with partial contact of surfaces.

First, we calculate the changes of deformation energy during variations of the opening region D and the loads.

Let's suppose that a solid V_o contains a crack Ω and loads are applied to the surface Σ_o of the solid.

If we consider two elastic states $\sigma_{ik}^{(1)}, \varepsilon_{ik}^{(1)}$ and $\sigma_{ik}^{(2)}, \varepsilon_{ik}^{(2)}$, the change in elastic energy is equal to

$$\Delta U \equiv U_2 - U_1 = 1/2 \int\limits_{V_o} (\sigma_{ik}^{(2)}\varepsilon_{ik}^{(2)} - \sigma_{ik}^{(1)}\varepsilon_{ik}^{(1)})dV, \tag{4.1}$$

where U_i denotes the elastic energy in the i-th state. It can be shown that (4.1) may be represented in the following form

$$\Delta U = 1/2 \int\limits_{\Omega} (\sigma_{33}^{(2)} - \sigma_{33}^{(1)})(u_3^{(2)} + u_3^{(1)})dx_1\, dx_2. \tag{4.2}$$

The final result (4.2) does not include integrals over the volume V_o and over the surface Σ_o, so (4.3) may be applied also to a crack in an unbounded medium.

52

We now consider equilibrium states from the set K. For any one of them (with opening region $D_K^{(1)}$) we have $D_K^{(1)} \subset D$ and obtain a sequence of equilibrium states from the set K with opening regions $D_K^{(1)} \subset D_K^{(2)} \subset \cdots \subset D$ and

$$\bigcup_{i=1}^{\infty} D_K^{(i)} = D \tag{4.3}$$

according to section 3. We shall show that the elastic energy increases during the transition from the i-th equilibrium state with the opening region $D_K^{(i)}$ to the $(i+1)$-th with the opening region $D_K^{(i+1)}$. Let us remind (see section 3) that the transition from the i-th state to the $(i+1)$-th is performed by "removing" the loads in the subdomain $F_K^{(i)'} \subset F_K^{(i)} = \Omega \setminus D_K^{(i)}$, where $\sigma_{33}^{(i)} > \sigma_{33}^o$. Denoting $F_K^{(i)'} = \delta D_K^{(i)}$ we obtain from (4.3) the increment of elastic energy $\Delta U^{(i)}$ during the transition from the i-th to the $(i+1)$-th equilibrium state

$$\Delta U^{(i)} = \frac{1}{2} \int\limits_{\delta D_K^{(i)}} \delta W^{(i)}(\sigma_{33}^o - \sigma_{33}^{(i)})n_3^+ dS. \tag{4.4}$$

Since $(\sigma_{33}^o - \sigma_{33}^{(i)})n_3^+ > 0$, $\delta W^{(i)} \geq 0$, $(x_1, x_2) \in \delta D_K^{(i)}$, we have $\Delta U^{(i)} \geq 0$. For the sequence of equilibrium states from the set K with the opening regions $D_K^{(1)} \subset D_K^{(2)} \subset \cdots \subset \cdots \subset D$ we have $U_1 \leq U_2 \leq \cdots \leq U_n \leq \cdots \leq U$, where U is the elastic energy corresponding to the equilibrium state with the opening region D, i.e. the solution of the original problem (1.1)-(1.4).

Using formula (4.2), it can be shown that the elastic energy reaches its maximum in the equilibrium state (correspondingly, the total potential energy of the system admits its minimum). Indeed, let us consider the following two states:

1. σ_{33}, W corresponding to the solution of the problem (1.1)-(1.4) and
2. $\sigma_{33}^{(1)}, u_3^{(1)}$ which is one of the states belonging to the set K.

We rewrite (4.2)

$$\Delta U = -\frac{1}{2} \int\limits_{\Omega} (u_3^{(1)} - W)(\sigma_{33}^{(1)} - \sigma_{33})n_3^+ dS + \int\limits_{\Omega} u_3^{(1)}(\sigma_{33}^{(1)} - \sigma_{33})n_3^+ dS, \tag{4.5}$$

where σ_{33}, W are the solutions of problem (1.1)-(1.4). The first term in (4.5) is obviously negative. Let's consider the second term. Since $u_3^{(1)} \geq 0$, $\sigma_{33} \leq \sigma_{33}^o$, $n_3^+ = -1$, the second term is also negative. Thus $\Delta U \leq 0$, and the elastic energy is maximal for the solution of problem (1.1)-(1.4). We note that the maximality of the elastic energy is established here in the general case of a solid bounded by the surface Σ_o, since (4.2) was obtained for such solids [3].

Thus, we have proved the following statement.

Statement 5. Let an elastic medium with crack Ω and prescribed loads be given. Then, among all equilibrium states from the set, the state for which the elastic energy reaches its maximum (the system's potential energy reaches its minimum) is the solution of problem (1.1)-(1.4).

We arrived at the variational principle that coincides with the one established in [3] considered for the same set.

The proof of Statement 5 demonstrates the equivalence of two approaches to construct the solution for the problem of a cavity (a crack-cut) with contact regions: The one proposed in the present work (section 3) and the one based on the extremality property of elastic energy U are equivalent.

We now show that both approaches lead to the construction of a solution, which is nonsingular near the boundary of the contact region.

5 Asymptotics of the solution near the contact contour

We now consider the behaviour of the solution of the problem of a cavity (a crack) with partial contact of the surfaces near the boundary separating the regions of contact and opening, depending on the constraints in these regions (1.2), (1.3) and on the geometry of initial opening

$$u(x_1, x_2).$$

To analyse the asymptotics of the solution near an arbitrary smoothness point of the boundary Γ separating the contact region F and the opening region D, we introduce the local coordinate system XYZ, where the Z axis is tangential to the boundary of the region D in the given point, the Y axis is perpendicular to the plane $x_3 = 0$, and the X axis is directed such that $x \leq 0$ corresponds to the opening region D.

From (1.1)-(1.4) it follows that

$$u_y^+ - u_y^- = -u(x) \,, x \geq 0 \tag{5.1}$$

$$\sigma_{yy}^{\pm}(x,0) = \sigma(x) \,, x \leq 0, y \to \pm 0,$$

where $\sigma(x)$ is determined by external loads, and $u(x)$ is the initial opening of the cavity.

We shall assume that $u(x), \sigma(x)$ satisfy a Hölder condition [10,11] and are analytic in a neighoourhood of the point $x = 0$. From the symmetry of the original problem relative to the plane $y = 0$ it follows that

$$\sigma_{zy}^{\pm} = \sigma_{zx}^{\pm} = 0, \quad |x| < \infty, y = 0.$$

In the local coordinate system the stress state is described by the Kolosov-Muskhelishvili formulae (μ-shear modulus)

$$\sigma_{xx} + \sigma_{yy} = 2[\Phi(\eta) + \Phi^*(\eta)], \quad \eta = x + iy$$

in the case of plane strain

$$\sigma_{yy} - i\sigma_{xy} = \Phi(\eta) + \Omega(\eta^*) + (\eta - \eta^*)\Phi'^*(\eta) \tag{5.2}$$

$$2\mu(u_x' + iu_y') = x\Phi(\eta) - \Omega(\eta^*) - (\eta - \eta^*)\Phi'^*(\eta)$$

$$u_\alpha' = \partial u_\alpha/\partial x, \alpha = x, y; \mu = E/2(1 + \nu), x = 3 - 4\nu$$

Substituting the representations (5.2) into (5.1), we come to the following conjugation problem [10]

$$\Phi^+ - \Phi^- = ig_o'(x), x \geq 0, \quad \Phi^+ + \Phi^- = \sigma(x), x \leq 0$$

$$g_o = 2\mu(u_y^+ - u_y^-)/(x + 1) = -2\mu u/(x + 1) \tag{5.3}$$

$$g_o' = dg_o/dx.$$

We consider the canonical solution of the problem (5.3), $x_o = \eta^{1/2}$, for which $x_o^+ = x_o^-; x \geq 0; x_o^+ = -x_o^-, x \leq 0, y = 0$.
The solution of the boundary-value problem (5.3) in the class of functions bounded at infinity and unbounded at the point $\eta = 0$, has the form [11]

$$\Phi(\eta) = (M(\eta) + M_1(\eta) + B)/x_o(\eta)$$

$$M(\eta) = \frac{1}{2\pi i} \int\limits_{-\infty}^{o} \frac{\sigma(t)x_o^+(t)dt}{t - \eta}, \quad M_1(\eta) = \frac{1}{2\pi i} \int\limits_{o}^{\infty} \frac{ig_o'(t)x_o^+(t)dt}{t - \eta}$$

(B is a constant determined by the conditions at infinity).
To estimate $M(\eta)$, $M_1(\eta)$, we consider the functions $w(\eta) = \sigma(\eta)x_o(\eta)$, $w_1(\eta) = ig_o'(\eta)x_1(\eta)$ where $x_1(\eta) = \eta^{1/2}$, $\sqrt{\eta} = \pm\sqrt{x}$, $x \geq 0, y \to \pm 0$. A cut is drawn in the XY plane along the positive half-axis. We obtain $(M - w)^+ = (M - w)^-$ and $(M_1 - w_1)^+ = (M_1 - w_1)^-$ for $y \to \pm 0$. Thus, the

functions $M(\eta) - w(\eta)$ and $M_1(\eta) - w_1(\eta)$ are analytic in a neighbourhood of the point $\eta = 0$; $M(\eta) - w(\eta) + M_1(\eta) - w_1(\eta)$ is an analytic function.
Thus, the function $\Phi(\eta)$ has the following representation

$$\Phi(\eta) = \sigma(\eta) + ig'_o(\eta)x_1(\eta)x_o^{-1}(\eta) + R_n(\eta)x_o^{-1}(\eta)$$

$$R_n(\eta) = \sum_{k=1}^{n+1} A_k \eta^{k-1}(k - 1/2) \tag{5.4}$$

near the point $\eta = 0$.
After incorporating (5.4) and (5.2) it follows from the solution of the boundary-value problem (5.1) that the distribution of stresses and displacements along the x-axis has the form

$$\sigma_{yy}^{\pm}(x,0) = \sigma(x), \ x \leq 0$$

$$\sigma_{yy}^{\pm}(x,0) = \sigma(x) + (1/2)x^{-1/2}A_1 + \cdots + (n - 1/2)x^{n-3/2}A_n, \ x \geq 0 \tag{5.5}$$

$$\sigma_{yy}(x,0) \leq \sigma(x), \ x \geq 0; \ u_y(x,0) = -u(x), \ x \geq 0$$

$$u_y(x,0) = -u(x) + 2(1 - \nu)\mu^{-1}(r^{1/2}A_1 + \cdots + (-1)^{n+1}r^{n-1/2}A_n), \ x \leq 0,$$

$$r = |x|$$

in a neighbourhood of the point $x = 0$. The function $u(x)$, defined for $x \leq 0$, is analytically continued to $x \geq 0$. The distributions of stresses and displacements along the x-axis are related through the coefficients A_k. We shall analyse the asymptotics of (5.5) for various cavity openings.
Let the equation of the cavity surface have the form $u_o = |x|^\alpha \ F_p(x) + B$, $x \leq 0$, where $F_p(x)$ is a polynomial of degree $p, 0 < \alpha \leq 1$ (in general, the surface is piecewise continuous in the point $x = 0$). Since $x \geq 0$ corresponds to the contact region and to the opening region of the cavity,

$$u_y(x) \geq u_o(x), \ \ x \leq 0, y \to \pm 0 \tag{5.6}$$

$$\sigma_{yy} \leq \sigma(x), \ \ x \geq 0, y \to \pm 0. \tag{5.7}$$

Let's assume that $A_1 \neq 0$. From (5.5) and (5.7) it follows that $A_1 < 0$.
Since the initial opening $u(x)$ of the cavity for $x \geq 0$ is an analytic function in the neighbourhood of the point $x = 0$, the asymptotics of the function $u(x)$ has the form $u(x) = Cr^\gamma + B$, where γ is an integer, $C = const, r = |x|$. Substituting this expression into (5.6), (5.5) and assuming that $F_p = Dr^p \ D = const > 0$ we obtain the inequality

$$-Cr^\gamma + A_1\beta r^{1/2} - B > -B - Dr^{p+\alpha}, r \simeq 0.$$

From this inequality it follows necessarily that $p = 0$, $0 < \alpha < 1/2$ for $A_1 < 0$.

Thus, the case $A_1 < 0$ is possible only provided that $p = 0$, $0 < \alpha \leq 1/2$ for a piecewise smooth cavity surface.

If $p \neq 0$ or $p = 0$, $1/2 < \alpha \leq 1$, it is necessary that $A_1 = 0$. Then it follows from condition (5.7) that $A_2 < 0$, the distribution of stresses in the contact region being non-singular.

For $p = 0$, $0 < \alpha < 1/2$ and arbitrary finite A_1 (which corresponds to finite values of compression stresses) a complete closing of the cavity near the angle point is impossible: In this case for a fixed boundary of the region the "overlapping" of cavity surfaces does not occur, which would make the formation of contact zones possible. This means that it is possible to determine a priori, without solving the problem, the regions of cavity surfaces which will not meet to form a contact (judging from the geometry of surfaces). The stresses are compressing and singular near the corresponding region, where the surfaces form an angle . A similar analysis is done for crack-cuts.

We note that the established dependence of the conditions for the contact region formation on parameters corresponds to the results of numerical calculations for cavities with elliptic cross-sections [4]: For an initial opening $u(x_1, x_2) = b(1 - x_1^2/a^2 - x_2^2/b^2)^{\alpha/2}$ of the cavity the formation of the contact region begins at the cavity's boundary (in plane $x_3 = 0$) if $\alpha \geq 1$, and inside the cavity if $0 < \alpha < 1$.

Remark. In the boundary-value problem (1.1)-(1.4) for the cavity with cross-section Ω and initial opening $u(x_1, x_2)$, $(x_1, x_2) \in \Omega$, the boundary condition in the cavity opening region $D = \Omega \setminus F$ coincides with the boundary conditions of the problem for a crack-cut occupying the region D in the elastic medium. The difference consists in an additional requirement, which holds in the crack problem for the value of the stress intensity factor that is required to be zero at the boundary of D. This makes it possible to use the class of the solutions for the problem of equilibrium of crack-cuts to construct the solutions of the problem of equilibrium of crack-cuts with contact regions. The equation defining the boundary Γ between contact and opening regions in the problem of crack-cuts occupying the region D is the condition of searching for the contour of a crack-cut (region D) for which $A_1(x_1^o, x_2^o) = 0$, $(x_1^o, x_2^o) \in \Gamma$.

The regular algorithm discussed in section 3 leads us to the construction of a

Conclusions: In the present paper a number of properties of the solution of the normal problem is established. On their basis a new regular algorithm is suggested for constructing the solution of the problems with contact and opening regions. The equivalence between the known approaches to the problem of cavities (crack-cuts) with contact regions and the proposed approach is demonstrated. Some advantanges of the proposed approach are discussed.

References

[1] Goldstein, R.V., Spector, A.A.: A variational method for the investigation of thre–dimensional elasticity problems on a plane crack with sliping and sticking of crack surfaces. Prikladnaya Matematika i Mekhanika, No.2, 276-285 (1983)(in Russian).

[2] Goldstein, R.V., Entov, V.M.: Variational estimates of the stress intensity factor on the contour of a plane crack of the normal opening. Izvestiya AN SSSR. Mekhanika Tverdogo Tela, No.4, 59-64 (1975)(in Russian).

[3] Goldstein, R.V., Spector, A.A.: Variational estimates of solutions of some thre–dimensional elasticity problems with unknown boundary. Izvestiya AN SSSR. Mekhanika Tverdogo Tela, No.2, 82-94 (1978)(in Russian).

[4] Balueva, A.V., Goldstein, R.V., Zazovskiǐ, A.F.: A method for the calculation of surface displacements of cavities in three–dimensional media. Fiziko-Tekhnicheskie problemy razrabotki poleznykh iskopaemykh, No.6, 3-6 (1984)(in Russian).

[5] Nikishin, V.S., Shapiro, G.S.: On the local axysymmetrical compression of an elastic layer weakened by annular and circular cracks. Prikladnaya Matematika i Mekhanika, No.1, 139-144 (1974)(in Russian).

[6] Mossakovskiǐ, V.I., Zagubezhenko, P.A.: On the compression of elastic isotropic planes weakened by rectilinear cracks. Doklady AN Ukr. SSR. Ser.A., No.5, 385-390 (1954)(in Russian).

[7] Boiko, L.T., Berkovich, P.E.: The contact problem for the plane with crack of changing width. Prikladnaya Matematika i Mekhanika, No.6, 1084-1089 (1974)(in Russian).

[8] Lur'e, A.I.: Theory of the elasticity. Nauka, Moscow 1970. (in Russian).

[7] Boiko, L.T., Berkovich, P.E.: The contact problem for the plane with crack of changing width. Prikladnaya Matematika i Mekhanika, No.6, 1084-1089 (1974)(in Russian).

[8] Lur'e, A.I.: Theory of the elasticity. Nauka, Moscow 1970. (in Russian).

[9] Kerchman, V.I.: Extremal properties of elasticity energy and new variational principles in unilateral problems on punches and cracks. Izvestiya. AN SSSR. Mekhanika Tverdogo Tela, No.5, 68-77 (1981)(in Russian).

[10] Muskhelishvili, N.I. Some basic problems in the mathematical theory of elasticity, Noordhoff, Groningen, Holland, 1953.

[11] Muskhelishvili, N.I. Singular integral equations, Ed. by J.R.M.Radock, Noordhoff, Groningen, Holland, 1953.

Robert Goldstein, Yuriĭ Zhitnikov
Institute for Problems in Mechanics
Russian Academy of Sciences
101 Prospect Vernadskogo
117526 Moscow, Russia
e-mail: goldst@ipm.msk.su

u–Factorization and Toeplitz Operators with Infinite Index

Sergeĭ M. Grudskiĭ

This work is devoted to the theory of normal solvability (Fredholm and semi–Fredholm properties, one–sided and generalized invertibility) of Toeplitz operators in L_p spaces on the unit circle.

The symbols $a(t)$ of the operators we consider are supposed to have discontinuities of the second kind of infinite index type, which means that $\arg a(t)$ gets an infinite increment in a neighborhood of the discontinuity. Problems of such kind have been studied since the late sixties and much progress has been made in the cases of almost–periodic and semi–almost–periodic discontinuities, weakly oscillating and multiplicatively–periodic symbols (see [1]–[10]).

The basic method used in this work is a generalized factorization (u–factorization) of so–called u–periodic functions ([11]– [13]), i.e. functions representable as a superposition $a(u(t))$ where $u(t)$ is an inner function; recall that an analytic and bounded function $u(z)$ on the unit circle is called an inner function if $|u(t)| = 1$ for almost all $|t| = 1$. Note that if $u(x) = e^{ix}, x \in \mathbb{R}$, then $a(e^{ix})$ is a usual 2π–periodic function on the real axis $\mathbb{R}$. This method allows us to investigate wide classes of problems with infinite index. In particular these classes contain not only cases of functions with very quickly growing argument $\arg a(t)$ ((2.19)–(2.20)) but (in contrast to [12]–[14]) also functions whose argument $\arg a(t)$ approaches infinity very slowly ((2.22)–(2.23)). It should also be emphasised that the technique proposed is well adapted to matrix case and thus enables us to investigate matrix problems with infinite index almost in the same degree of generality as their scalar counterparts (Theorem 3.1). It is worth mentioning that the theory of matrix operators of such kind is much less advanced than the corresponding theory for the scalar case. Only a few papers devoted to some classes of operators with almost-periodic matrix symbols can be referred to in this context [14].

Let us introduce the main idea of the u-factorization method. It is well–known [15] that any inner function $u(t)$ can be represented as a product $u(t) = B(t)E(t)$ where

$$B(t) = \prod_{j=-\infty}^{\infty} \frac{\bar{z}_j}{|z_j|} \cdot \frac{z_j - t}{1 - \bar{z}_j t} \tag{0.1}$$

is a Blaschke product with the zero set $\{z_j\}_{j=-\infty}^{\infty}$ (so that $|z_j| < 1$ and $\sum(1 -$

This work was performed under financial support of Russian fund of fundamental investigations (grant 93–011–28).

60

$|z_j|) < \infty)$ and

$$E(t) = \exp\left\{-\int_{-\pi}^{\pi} \frac{e^{i\theta} + t}{e^{i\theta} - t} d\mu(\theta)\right\} \tag{0.2}$$

is a singular inner function generated by a positive singular measure $\mu(\theta)$.

It turns out [15] that every inner function can be analytically continued onto the complement of the set $M = (\{\bar{z}_j^{-1}\}_{j=-\infty}^{\infty} \cup \mathcal{U})$, whose Lebesgue measure equals zero; here $\mathcal{U}$ is the closure of the support of the singular measure $\mu(t)$. Furthermore, if $|z| < 1$ then $|u(z)| < 1$ and if $|z| > 1$ then $|u(z)| > 1$.

From what was said in the preceding paragraph it follows that if the function $a(t)$ is analytic for $|z| < 1$ or for $|z| > 1$, then the superposition $a(u(t))$ is also analytic in the corresponding region. Consequently, the superposition introduced above transforms many conventional factorizations of Wiener–Hopf type into so–called generalized factorizations [16], which allows us to investigate Toeplitz operators with symbols of the form $a(u(t))$. Wide classes of discontinuities can be modeled with the help of this superposition. For example, if $a(t)$ is continuous then $a(u(t))$ can have one, several or a countable set of discontinuities of second kind of oscillating type.

In Section 1 we briefly present some general results on u-factorization of matrix-valued functions. Section 2 is devoted to a more detailed investigation of some scalar problems with infinite index, while Sections 3 is concerned with such problems in the matrix case.

1 Factorization of u–periodic matrix–functions.

Denote by $L_p^{(n)}$ and $L_p^{(n \times n)}$ the spaces of vector–functions and matrix–functions of order n, respectively, whose elements are the functions summable in the pth power on the unit circle Γ_0. Let S be the Cauchy singular integral operator on L_p $(1 < p < \infty)$, where here and in what follows the superscript n will be omitted for $n = 1$.

The standard singular projectors $P^{\pm}$ are given by $P^{\pm} = \frac{1}{2}(I \pm S)$ on L_p and extended to $L_p^{(n)}$ in the natural way. Put $L_p^{\pm(n)} = P^{\pm}(L_p^{(n)})$ and $L_p^{\pm(n \times n)} = P^{\pm}(L_p^{(n \times n)})$ and denote the Hardy classes of vector and matrix functions which are analytic and bounded inside or outside Γ_0 by $H_{\infty}^{\pm(n)}$ and $H_{\infty}^{\pm(n \times n)}$, respectively.

The main object of our interest is Toeplitz operators

$$T(A) = P^+ A(t)I, \quad A(t) \in L_{\infty}^{(n \times n)}, \tag{1.1}$$

on $L_p^{+(n)}$, where $1 < p < \infty$. Note that $A(t)$ is usually called the symbol of the corresponding Toeplitz operator.

A matrix–function $A(t)$ $(\in L_\infty^{(n\times n)})$ is said to allow a p–factorization if it can be represented as

$$A(t) = A_-(t)D(t)A_+(t), \tag{1.2}$$

where $D(t) = \operatorname{diag}[t^{\kappa_1}, t^{\kappa_2}, ..., t^{\kappa_n}]$, κ_j are integers $(\kappa_1 \geq \kappa_2 \geq ... \geq \kappa_n)$, called the partial indices of the factorization, and the factors $A_\pm(t)$ satisfy the following conditions:

$\alpha)$ $A_+(t) \in L_q^{+(n\times n)}$, $A_+^{-1}(t) \in L_p^{+(n\times n)}$, $A_-(t) \in L_p^{-(n\times n)} \oplus \{c\}$, $A_-^{-1}(t) \in L_q^{-(n\times n)} \oplus \{c\}$, where $p^{-1} + q^{-1} = 1$, and $\{c\}$ is the space of constant matrix–functions;

$\beta)$ the operator $K = A_-(t)SA_-^{-1}I$ is bounded on $L_p^{(n)}$.

We shall say that (1.2) is a standard factorization if

$$A_+^{\pm 1}(t) \in H_\infty^{+(n\times n)}, \quad A_-^{\pm 1} \in H_\infty^{-(n\times n)}. \tag{1.3}$$

Last but not least, $A(t)$ $(\in L_\infty^{(n\times n)})$ is said to admit a (p, u)–factorization if it is representable in the form

$$A(t) = A_-(t)D(u(t))A_+(t), \tag{1.4}$$

where $D(u(t)) = \operatorname{diag}[u^{\kappa_1}(t), u^{\kappa_2}(t), ..., u^{\kappa_n}(t)]$ and the factors $A_\pm(t)$ are subject to the conditions $\alpha)$ and $\beta)$ a (p, u)–factorization for which (1.3) is in force is referred to as a standard u–factorization.

It is well known [17] that the existence of a p–factorization of the symbol $A(t)$ is equivalent to the Fredholmness of the operator (1.1). The following theorem shows the strong relation between the existence of a (p, u)–factorization and the normal solvability of the operator.

Theorem 1.1 *Suppose the matrix–function $A(t) \in L_\infty^{(n\times n)}$ admits a (p, u)–factorization for some inner function $u(t)$. Then $T(A)$ is a generalized–invertible operator on $L_p^{+(n)}$. Furthermore, if $\kappa_j \geq 0$ for all j then $T(A)$ is left–sided invertible, whereas if $\kappa_j \leq 0$ for all j then $T(A)$ is right–sided invertible; if $\kappa_j = 0$ for all j then $T(A)$ is two–sided invertible. In either case a generalized inverse is given by*

$$[T(A)]^{-1} = P^+ A_+^{-1} D^{-1}(u(t)) P^+ A_-^{-1} I.$$

We now associate with each inner function $u(t)$ a map γ_u by the formula

$$(\gamma_u F)(t) = F(u(t)). \tag{1.5}$$

Theorem 1.2
1) The map γ_u is a homomorphism of the algebra $L_\infty^{n\times n}$ into itself, and it leaves invariant the subalgebras $H_\infty^{\pm(n\times n)}$.

2) The map γ_u is a linear bounded operator on $L_p^{(n\times n)}$, $1 < p < \infty$, and the subpaces $L_p^{\pm(n\times n)}$ are invariant with respect to γ_u. Moreover, for every matrix–function $F(t) \in L_p^{(n\times n)}$ we have the estimates

$$\left(\frac{1 - |u(0)|}{1 + |u(0)|}\right)^{\frac{1}{p}} \|F\|_{p,n} \le \|\gamma_u F\|_{p,n} \le \left(\frac{1 + |u(0)|}{1 - |u(0)|}\right)^{\frac{1}{p}} \|F\|_{p,n}, \qquad (1.6)$$

where the norm of a matrix–function $F(t) = \{f_{k,j}(t)\}_{k,j=1}^{n}$ is defined by

$$\|F\|_{p,n} = \left(\sum_{j=1}^{n}\sum_{k=1}^{n}\left(\int_{\Gamma_0} |f_{k,j}(t)|^p |dt|\right)^{\frac{2}{p}}\right)^{\frac{1}{2}}. \qquad (1.7)$$

The previous theorem is our main tool in constructing u–factorizations. Here, for example, is an immediate consequence of this theorem.

Theorem 1.3 *If the matrix–function $A(t)$ allows a standard factorization, then $A(u(t))$ admits a standard u–factorization for each inner function $u(t)$.*

Many conventional classes of matrix–functions posses standard factorizations. For instance, if $A(t)$ belongs to a Hölder space or if $A(t)$ can be expanded into an absolutely converging Fourier series, then Theorem 1.3 gives a standard u–factorization of the superposition $A(u(t))$ under the condition that $\det A(t) \ne 0$. It looks quite natural to try generalizing Theorem 1.2 to the case of arbitrary p–factorizable matrix–functions $A(t)$. Indeed, because of statement 2) of Theorem 1.2, the conditions $\alpha)$ are always satisfied for the factors $A_{\pm}(u(t))$; however, we have not been able to prove the validity of condition $\beta)$ for arbitrary inner functions. So we shall in the rest of this section restrict ourselves to the well–known class of sectorial functions (see [10] and [17]).

We denote by $G_p^{(n\times n)}$ the class of all matrix–functions $A(t) \in L_\infty^{(n\times n)}$ which have a representation

$$A(t) = h^-(t)D(t)A_0(t)h^+(t)$$

where $[h^-(t)]^{\pm 1} \in H_\infty^{-(n\times n)}$, $[h^+(t)]^{\pm 1} \in H_\infty^{+(n\times n)}$, $D(t)$ is the same as in (1.2), and $A_0(t)$ is a unitary matrix for almost all $t \in \Gamma_0$ whose spectrum is contained in the sector

$$|\arg z| \le \pi/\max(p,q), \quad p^{-1} + q^{-1} = 1. \qquad (1.8)$$

Theorem 1.4 *Assume $A(t) \in G_p^{(n\times n)}$ and $r \in [\min(p,q); \max(p,q)]$. Then the Toeplitz operator $T(A(u(t)))$ is two–sided, left–sided or right–sided invertible on $L_r^{+(n)}$ in dependence on whether all $\kappa_j = 0$, all $\kappa_j \ge 0$ or all $\kappa_j \le 0$. In either case a generalized inverse is given by*

$$[T(A(u(t)))]^{-1} = [h^+(u(t))A_{0+}(u(t))]^{-1}P^+[A_{0-}(u(t))D(u(t))h^-(u(t))]^{-1}I,$$

where $A_0(t) = A_{0-}(t)A_{0+}(t)$ is an r–factorization of the matrix–function $A_0(t)$.
The proof of this theorem is based on the observation that the matrix–function $A_0(t)$ is also unitary–valued for almost all $t \in \Gamma_0$ and that its spectrum is located in the sector (1.8).

The class $G_p^{(n \times n)}$ is sufficiently large. For example, it includes all 2–factorizable matrix–functions when $p = 2$ and all continuous nonsingular function when $n = 1$ and $1 < p < \infty$ [18].

2 Problems with infinite index. Scalar case.

Everywhere below we shall assume that the symbol $a(t)$ of the Toeplitz operator has a unique discontinuity, at the point $t = 1$, say. We henceforth write $t = e^{i\theta}$ with $\theta \in [-\pi, \pi]$. Let $f(\theta)$ be a real–valued function which is defined and continuous on the set $[-\pi, 0) \cup (0, \pi]$ and increases monotonously on each of these two semi–intervals such that $\lim_{\theta \to \pm 0} f(\theta) = \mp\infty$. Without loss of generality we assume that $f(-\pi) = f(\pi) = 0$. We introduce inverse flipped function $\theta(x) = f^{-1}(-x)$, which is defined for all $x \in \mathbb{R}\backslash\{0\}$.

We put $\Delta(n) = (\theta(n) - \theta(n + 1))$, $n \in \mathbb{Z}$, and introduce the sequence of functions

$$\psi_n(\varepsilon) = \frac{\theta(n) - \theta(n + \varepsilon)}{\Delta(n)}, \ \varepsilon \in [-1/2; 1/2], \ n = \pm 1, \pm 2, \ldots$$

Theorem 2.1 *Assume that the functions $a(t) \in L_\infty$ is invertible in L_∞ and continuous on $\Gamma_0\backslash\{1\}$. At the point $t = 1$, assume that*

$$\lim_{\theta \to 0}(\arg a(\exp(i\theta)) - 2\pi\delta f(\theta)) = 0, \ \delta = \pm 1,$$

where the function $\theta(x) = f^{-1}(-x)$ satisfies the following conditions:

1) $\sup\limits_{n \in \mathbb{Z}} \sum\limits_{\substack{j = -\infty \\ j \neq n}}^{\infty} \left(\frac{\Delta(j)}{\theta(j) - \theta(n)}\right)^2 < \infty$; 2) $\inf\limits_{n \in \mathbb{Z}} \frac{\Delta(n + 1)}{\Delta(n)} = \Delta > 0$;

3) as $n \to \pm\infty$, the sequence $\{\psi_n(\varepsilon)\}_{-\infty}^{\infty}$ converges uniformly on $[-1/2; 1/2]$ to some continuous function $\psi(\varepsilon)$ satisfying $\psi(1/2) > 0$ and $\psi(-1/2) < 0$.

Then for each $p \in (1, \infty)$ there exists a sufficiently large $M > 0$ such that $a(t)$ admits a (p, B_M)–factorization of the form

$$a(t) = g(t)B_M^\delta(t), \tag{2.1}$$

where $g(t)$ allows a p–factorization and B_M is a Blaschke product as in (0.1) with $z_k = r_k \exp(i\theta(k))$ and $r_k \in (0, 1)$ given by $((1 - r_k)/(1 + r_k)) = \Delta(k)/M$, $k = \pm 1, \pm 2, \ldots$

64

Proof We only consider the case $\delta = 1$. Choose M so large that the numbers r_k given by $((1 - r_k)/(1 + r_k)) = \Delta(k)/M$, $k = \pm 1, \pm 2, \ldots$ lie between 0 and 1, define B_M as in the theorem and put $f_M(x) = \arg B_M(e^{i\theta(x)})$. Suppose $x > 0$ and $x = n + \varepsilon$, where $n \in \mathbb{Z}$, and $\varepsilon \in [-1/2; 1/2]$. With the help of elementary geometrical arguments [11] it is easy to calculate that

$$f_M(x) = -2[\pi(n - 1) + \sum_{k \neq n} \varphi_k(\theta(x)) + \tilde{\varphi}_n(\theta(x))], \qquad (2.2)$$

where $\varphi_k(\theta(x)) = \arctg[\frac{\Delta(k)}{M}\ctg\frac{\theta(x) - \theta(k)}{2}]$ and

$$\tilde{\varphi}_n(x) = \left\{ \begin{array}{ll} \varphi_n(\theta(x)), & \varepsilon \leq 0, \\ \pi + \varphi_n(\theta(x)), & \varepsilon > 0. \end{array} \right.$$

Let us consider the terms on the right side of equality (2.2). Everywhere below the functions $O_j(x)$ will satisfy an inequality of the form

$$|\theta_j(x)| \leq \text{const} \cdot M^{-1}. \qquad (2.3)$$

At first we observe that

$$\sum_{k=1}^{n-2} \varphi_k(\theta(x)) = \sum_{k=1}^{n-2} \arctg\frac{2}{M} \cdot \frac{\Delta(k)}{\theta(x) - \theta(k)} + O_1(x). \qquad (2.4)$$

Indeed from inequality

$$|\ctg u - u^{-1}| \leq \text{const}|u|, \ |u| < u_0, \qquad (2.5)$$

where $u_0 < \pi$, it follows that

$$|O_1(x)| = \left| \sum_{k=1}^{n-2} \left(\arctg\left[\frac{\Delta(k)}{M}\ctg\frac{\theta(x) - \theta(k)}{2}\right] - \arctg\frac{2}{M} \cdot \frac{\Delta(k)}{\theta(x) - \theta(k)} \right) \right| \leq$$

$$\leq \text{const} \left| \sum_{k=1}^{n-2} \arctg\frac{\Delta(k)}{M} \cdot \frac{\theta(x) - \theta(k)}{2} \right| \leq \frac{\text{const}}{M} \sum_{k=1}^{n-2} \Delta(k) \leq$$

$$\leq \frac{\text{const}}{M}(\theta(1) - \theta(n - 1)) \leq \text{const} \cdot M^{-1}.$$

We next prove that

$$\sum_{k=1}^{n-2} \arctg\frac{2}{M} \cdot \frac{\Delta(k)}{\theta(x) - \theta(k)} = \frac{2}{M} \sum_{k=1}^{n-2} \ln\left(1 + \frac{\Delta(k)}{\theta(x) - \theta(k)}\right) + O_2(x). \qquad (2.6)$$

This follows from the inequality

$$\left|\operatorname{arctg}\frac{2}{M}u - \frac{2}{M}\ln(1+u)\right| \le \frac{\text{const}}{M}u^2, \tag{2.7}$$

which is true for $u > u_0 > -1$, along with the estimation

$$\left|\sum_{k=1}^{n-2}\operatorname{arctg}\frac{2}{M}\cdot\frac{\Delta(k)}{\theta(x)-\theta(k)} - \frac{2}{M}\ln\left(1+\frac{\Delta(k)}{\theta(x)-\theta(k)}\right)\right| \le$$

$$\le \frac{\text{const}}{M}\sum_{k=1}^{n-2}\left(\frac{\Delta(k)}{\theta(x)-\theta(k)}\right)^2 \le \frac{\text{const}}{M},$$

where the last sum is bounded uniformly in n because of condition 1). Note that inequality (2.7) may be used in the case at hand since, owing to conditions 2) and 3) for $1 \le k \le n-2$,

$$\frac{\Delta(k)}{|\theta(x)-\theta(k)|} = \frac{1}{1+\frac{\theta(k+1)-\theta(x)}{\Delta(k)}} \le \frac{1}{\frac{\theta(k+1)-\theta(k+\frac{3}{2})}{\Delta(k)}} = \frac{1}{1+\frac{\Delta(k+1)}{\Delta(k)}\psi_n(1/2)} < u_0 < 1.$$

Calculating now the sum on the right side of equality (2.6) we get

$$\sum_{k=1}^{n-2}\varphi_k(\theta(x)) = \frac{2}{M}\ln\frac{\theta(x)-\theta(n-1)}{\theta(x)-\theta(1)} + O_3(x). \tag{2.8}$$

In a similar way we obtain

$$\sum_{k=n+2}^{\infty}\varphi_k(\theta(x)) = \frac{2}{M}\ln\frac{\theta(x)}{\theta(x)-\theta(n+2)} + O_4(x) \tag{2.9}$$

and

$$\sum_{k=-\infty}^{-1}\varphi_k(\theta(x)) = \frac{2}{M}\ln\frac{\theta(x)+\pi}{\theta(x)} + O_5(x). \tag{2.10}$$

The functions $\varphi_{n-1}(\theta(x))$ and $\varphi_{n+1}(\theta(x))$ can be proved to satisfy inequality (2.3) with the help of condition 3).
Now we pass to the term $\tilde{\varphi}_n(\theta(x))$. Using inequality (2.5) we get

$$\tilde{\varphi}_n(\theta(x)) = \left\{\begin{array}{ll} -\operatorname{arctg}2(M\psi_n(\varepsilon))^{-1} & , \ -1/2 \le \varepsilon \le 0 \\ \pi - \operatorname{arctg}2(M\psi_n(\varepsilon))^{-1} & , \ 0 < \varepsilon < 1/2 \end{array}\right\} + O_6(x).$$

As the sequence $\psi_n(\varepsilon))$ converges uniformly to a continuous function $\psi(\varepsilon)$, we have

$$\tilde{\varphi}_n(\theta(x)) = \alpha_0(\varepsilon) + c_1(x) + O_6(x), \tag{2.11}$$

where $\alpha_0(\varepsilon)$ is continuous function on $[-1/2; 1/2]$ of the kind

$$\alpha_0(\varepsilon) = \begin{cases} -\text{arctg}2(M\psi(\varepsilon))^{-1} & , \ -1/2 \leq \varepsilon \leq 0, \\ \pi - \text{arctg}2(M\psi(\varepsilon))^{-1} & , \ 0 < \varepsilon \leq 1/2, \end{cases}$$

and $c_1(x)$ is a function which is continuous for $x \in [n+1/2; n-1/2]$, $n = 1, 2, \ldots$ and satisfies

$$\lim_{x \to \infty} c_1(x) = 0. \tag{2.12}$$

Taking into account (2.8)–(2.11) we get

$$f_M(x) = -2 \left[\pi(n-1) + \frac{2}{M} \ln \frac{\theta(x) - \theta(n-1)}{\theta(x) - \theta(n+2)} + \alpha_0(\varepsilon) + c_1(x) + O_7(x) \right]. \tag{2.13}$$

Using conditions 2) and 3) one can prove that $\left| \ln \frac{\theta(x)-\theta(n-1)}{\theta(x)-\theta(n+2)} \right| < \text{const.}$
If $x = n + \varepsilon$ then $\varepsilon = \varepsilon(x) = x - [x + 1/2]$ and equality (2.13) transforms in

$$f_M(x) = -2\pi x + \alpha(\varepsilon(x)) + c_1(x) + O_8(x), \tag{2.14}$$

where $\alpha(\varepsilon) = -2(\alpha_0(\varepsilon) - \pi(1+\varepsilon))$ is a 1–periodic function which is continuous everywhere except at points of the type $x = n - 1/2$, $n \in N$. At these points $\alpha(\varepsilon)$ has discontinuities of the first kind. If the jump value at these points satisfies inequality (2.3) (uniformly in n), then the function $\alpha(\varepsilon(x))$ in equality (2.14) can be considered as a continuous 1–periodic function. Moreover, if necessary, we may slightly perturb the function $\alpha(\varepsilon(x))$ in order to achieve that the function $F(x) = -2\pi x + \alpha(\varepsilon(x))$ becomes monotonously decreasing. Proceeding in a similar way in the case $\kappa < 0$ one can show that (2.14) is true throughout the real axis $\mathbb{R}$.
Denoting the function inverse to $F(x)$ by $F^{-1}(y)$ we get

$$F^{-1}(y) = -\frac{y}{2\pi} + p(y), \tag{2.15}$$

where $p(y)$ is some 2π–periodical function which is continuous on $\mathbb{R}$. Acting by the map F^{-1} on both sides of equality (2.14) and taking into consideration uniform continuity we obtain

$$F^{-1}(f_M(x)) = x + c_2(x) + O_9(x), \tag{2.16}$$

where $c_2(x)$ is continuous on $\mathbb{R}$ and satisfies condition (2.12). Remembering that $x = -f(\theta)$ and $f_M(x) = \arg B_M(t)$, we conclude from (2.16) that

$$\exp(-2\pi i F^{-1}(-i \ln B_M(t))) = \exp(2\pi i f(\theta)) c_3(t) d(t),$$

where $c_3(t) = \exp(-2\pi i c_2(t))$, $d(t) = \exp(-2\pi i O_9(x))$.

Taking into consideration (2.15) we have

$$B_M(t)\exp(-2\pi i p(-i\ln B_M(t))) = a(t)c_4(t)d(t),$$

where $c_4(t)$ is continuous on Γ_0. We remark that the function $b(u) = e^{-2\pi i p(-i\ln u)}$ is continuous on Γ_0 and has index zero because $p(y)$ is 2π–periodic. So according to Theorem 1.4 (and the remarks to it) $b(B_M(t))$ allows a p–factorization. Finally notice that $d(t) = 1 + O_{10}(t)$ for M large enough (see (2.3)), which implies (2.1) with the function $g(t) = b(B_M(t))/(c_4(t)d(t))$. $\qquad\square$

Let us now provide some examples of functions $f(\theta)$ subject to the hypotheses conditions 1), 2), 3). We give the definition of these functions for $\theta > 0$ only:

$$f(\theta) = -c\theta^{-\lambda}\ln^\beta\theta^{-1}, \;\; c > 0, \;\lambda > 0, \;\beta \in I\!\!R; \tag{2.17}$$

$$f(\theta) = -c_1\theta^{-\lambda} + c_2\sin\theta^{-\beta}, \;\; c_1 > 0, \;\lambda > \beta > 0, \;c_2 \in I\!\!R; \tag{2.18}$$

$$f(\theta) = -c\exp(\theta^{-\lambda}), \;\; c > 0, \;\lambda > 0; \tag{2.19}$$

$$f(\theta) = -c\exp(\exp(\theta^{-\lambda})), \;\; c > 0, \;\lambda > 0; \tag{2.20}$$

$$f(\theta) = -c\ln^\lambda\theta^{-1}, \;\; c > 0, \;\lambda > 2. \tag{2.21}$$

In the last example, the parameter λ is bounded because condition 1) is not true for $0 < \lambda \le 2$.

The examples listed above demonstrate that the conditions of Theorem 2.1 are well adapted to rapidly increasing functions $f(\theta)$ but work worse for functions of slow growth. The following result may be applied to fairly wide classes of slowly increasing functions $f(\theta)$.

Theorem 2.2 *Suppose that all hypotheses of Theorem 2.1 are satisfied, except conditions 1) and 2). Instead of these conditions we require that*

$$1) \;\; \sup_{k\in Z}\frac{\theta(|k|+1)}{\theta(|k|)} = \Delta < 1; \;\; 2) \;\; \theta(k) = -c\theta(-k), \;\; c > 0.$$

Then for each $p \in (1,\infty)$ there exists a sufficiently large number $M > 0$ such that $a(t)$ admits a (p, B_M)–factorization (2.1), where $g(t)$ allows a p–factorization and the zeros of the Blaschke product $B_M(t)$ are of form $z_k = r_k\exp(i\theta(k))$ with the numbers r_k given by

$$\frac{1-r_k}{1+r_k} = \left\{\begin{array}{ll}\Delta(k)/M & , \;\; k = 1, 2, ..., \\ \Delta(-k-1)/M & , \;\; k = -1, -2, ...\end{array}\right.$$

The following example is typical for Theorem 2.2:

$$f(\theta) = -c\ln\theta^{-1}, \;\; c > 0. \tag{2.22}$$

Now we shall construct an example of more general character which illustrates the conditions of Theorem 2.2.

Assume that $\theta(k)$ is a monotonously decreasing sequence of numbers such that $\lim_{k\to\infty} \theta(k) = 0$, $\lim_{k\to\infty}(\Delta(k+1)/\Delta(k)) = \Delta$. Also let $\{\psi_n(\varepsilon)\}$ be a sequence of monotonous functions on $[-1/2; 1/2]$ satisfying $\psi_k(-1/2) = -1$, $\psi_k(0) = 0$, $\psi_k(1/2) = 1 - \Delta(k+1)/\Delta(k)$ and converging uniformly on $[-1/2; 1/2]$. Then the function

$$\theta(x) = \begin{cases} \theta(k) & , \ k \in \mathbb{N}, \\ \theta(k) - \psi_k(\varepsilon)\Delta(k) & , \ x = k + \varepsilon, \ \varepsilon \in [-1/2; 1/2[, \end{cases} \tag{2.23}$$

satisfies all conditions of Theorem 2.2. Notice that in the case $\Delta = 0$ the function $f(\theta)$ increases very slowly.

At the end of this section we formulate a theorem which shows that, under some extra assumptions, the function $g(t)$ in (2.1) can be represented in the form

$$g(t) = c(t)(1 + O_M(t)), \tag{2.24}$$

where $c(t)$ is continuous on Γ_0 and $O_M(t)$ satisfies inequality (2.3). This result is interesting by itself, but we shall need it in next section for matrix problems.

Theorem 2.3 *Assume that the function $a(t)$ is unimodular and continuous on $\Gamma_0\backslash\{1\}$ and has a discontinuity of second kind at the point $t = 1$ such that*

$$\lim_{\theta\to 0}(\arg a(\exp(i\theta)) - 2\pi\delta f(\theta)) = 0, \ \delta = \pm 1,$$

where $\theta(x) = f^{-1}(x)$ satisfies either the conditions of Theorem 2.1 or these of Theorem 2.2. Assume also that function $\psi(\varepsilon)$ in condition 3) allows the following asymptotic representation

$$\psi(\varepsilon) = c\varepsilon + O(\varepsilon^2), \ c > 0, \tag{2.25}$$

in a neighborhood of the point $\varepsilon = 0$.
Then for each $M > 0$ there exist an inner function $u_M(t)$ such that

$$a(t) = g(t)u_M(t) \tag{2.26}$$

with $g(t)$ of the form (2.24).

We remark that in the examples (2.17)–(2.21) we have $\psi(\varepsilon) = \varepsilon$, while in (2.22) we have $\psi(\varepsilon) = (1 - e^{-c\varepsilon})/(1 - e^{-c})$. So in either case condition (2.25) is satisfied.

3   Matrix case.

Let $A(t) = \{a_{ij}(t)\}_{i,j=1}^{n}$ be a matrix–function with entries of the form

$$a_{ij}(t) = c_{ij}(t)g_{ij}(\exp(2\pi i f(t))),$$

where c_{ij}, g_{ij} are continuous functions on Γ_0 and $f(t)$ is a real–valued continuous function on $\Gamma_0\backslash\{1\}$ with $\lim_{t\to 1\pm 0} f(t) = \mp\infty$. Associate the continuous matrix–function $A_0(t) = \{c_{ij}(1)g_{ij}(t)\}_{i,j=1}^{n}$ with $A(t)$.

Theorem 3.1 *Assume that* $\inf_{t\in\Gamma_0} |\det A(t)| > 0$ *and that the function* $a(t) = e^{(2\pi i f(t))}$ *allows the representation (2.26). Suppose also that the matrix–function* $A_0(t)$ *admits a standard factorization (1.2)–(1.3) with factors that are continuous on* Γ_0 *and with partial indices* κ_j, $1 \leq j \leq n$.
Then the operator $T(A)$ *is a* Φ_+*-operator,* Φ*-operator or* Φ_-*-operator on* $L_p^{+(n)}$ *if all* $\kappa_j \geq 0$, $\kappa_j = 0$ *or* $\kappa_j \leq 0$, *respectively.*

References

[1] Gohberg I., Feldman I.A. On Wiener–Hopf Integral Difference Equations, Soviet Mat. Dokl. , vol. 9 , 1312–1316 , 1968.

[2] Coburn L.A., Douglas R.G. Translation Operators on the Half–line, Proc. Nat. Acad. Sci. USA , vol. 62 , 1010–1013 , 1969 .

[3] Sarason D. Toeplitz Operators with Semi–almost Periodic Symbols, Duke Math. J. vol. 44, 2 , 357–364 , 1977 .

[4] Duduchava R.V., Saginashvili A.I. Convolution Integral Operators on a Half–line with Semi–almost Periodic Presymbol , Soobshch. Akad. Nauk Gruz. SSR ., vol 98, 1, 21–24, 1980.

[5] Xia J. The K–theory and the Invertibility of Almost Periodic Toeplitz Operators , Integr. Equat. and Oper. Theory , vol. 11, 2, 267–286, 1988.

[6] Abrahamse M.B. The Spectrum of a Toeplitz Operator with a Multiplicativly Periodic Symbol , J. Funct. Anal., vol. 31 , 2, 224–233, 1979.

[7] Power S.C. Fredholm Toeplitz Operators and Slow Oscillation, Can. J. Math., vol. 32, 5, 1058–1071, 1980.

[8] Monahov V.N., Semenko E.V. Classes of Correctness of Boundary Value Problems with Conjugation for Analytic Functions with Infinite Index, Dokl. AN SSSR, vol. 286, 1, 27–30. 1986.

[9] Grudskiĭ S.M. Singular Integral Equations and Rieman Boundary Value Problems with Infinite Index in $L_p(\Gamma,\omega)$, Izv, AN SSSR, mat., vol. 49, 1, 55–80, 1985.

[10] Böttcher A., Silbermann B. Analysis of Toeplitz Operators, Akademie-Verlag, Berlin and Springer-Verlag, Berlin, Heidelberg, New York, 1990.

[11] Grudskiĭ S.M. Singular Integral Operators with Infinite Index and Blaschke Products , Math. Nachr., vol. 129, 313–331, 1986.

[12] Grudskiĭ S.M. 180–184 Factorization of u–periodic Matrix–valued Functions and Problems with an Infinite Index , Soviet. Mat. Dokl., vol. 36, 1, 180–184, 1988.

[13] Grudskiĭ S.M. Matrix Singular Integral Operators with Infinite Index 2, Izv. vuzov, mat., 6, 69–72, 1991.

[14] Karlovich Yu.I., Spitkovskiĭ I.M. The Factorization Problem for Almost Periodic Matrix–functions and the Fredholm Theory of Toeplitz Operators with Semi-almost Periodic Matrix Symbols , in Linear and Complex Analysis Problem Book, Lect. Notes Math. , eds Havin V.P., Khrushchev S.V., and Nikolski N.K. , addr Springer–Verlag, Heidelberg , 1984 .

[15] Hoffman K. , Banach Spaces of Analytic Functions, Prentice–Hall, Inc., Englewood Cliffs, N.J. , 1962 .

[16] Spitkovskiĭ I.M. On the Vector–valued Riemann Boundary Problem with Infinite Defect Numbers and the Factorization of Matrix Functions Connected with This Problem , Mat. Sb. (N.S.), vol. 135, 533–550, 1988.

[17] Litvinchuk G.S., Spitkovskiĭ I.M. Factorization of Measurable Matrix Functions, Akademie–Verlag, Berlin and Birkhäuser Verlag, Basel, 1987.

[18] Krupnik N.Ya. Some Corollaries of the Hunt, Mackenhoupt and Wheeden Teorem, Matem. Issled., 47, 64–70, 1978.

Sergeĭ M. Grudskiĭ
Rostov State University
Rostov on Don
Ukraine

Implementational Details of the Boundary Element Method

Wolfgang Hackbusch

Abstract. We consider four different topics which appear in almost any implementation of a boundary element method (BEM). (1) For large problems the direct solution is no longer practical and has to be replaced by an iterative method. Here, we mention possible applications of the multi-grid method. (2) Full matrices are characteristic for BEM. Unfortunately, they lead to an amount of work and storage in the order $O(n^2)$. As a remedy we propose the panel clustering method. (3) Galerkin discretisations seem to be much more expensive because of the doubled number of integrations. We describe an implementation that is only by a certain factor more expensive than the usual collocation method. (4) Usually, the arising integrations cannot be computed exactly. Besides the regular and singular integrals, for which good quadrature techniques are known, there arise almost singular ones, where the singularity is close to the integration domain. We mention techniques for treating these integrals.

1. The Iterative Solution of the Systems of Linear Equations

For three-dimensional problems, the BEM leads to a relatively large system of linear equations. Since, in addition, a full matrix arises, direct methods like Gauß elimination or Cholesky are too expensive. One possible choice are the conjugate gradient method or one of its variants (cf. §9 in [Ha91a]). If the integral equation is of the second kind, the multi-grid method of the second kind is a very attractive method (§1.2). But also the standard multi-grid method can be applied to the hypersingular cases (§1.3).

1.1 General Construction of the Multi-Grid Method

The linear equations to be solved is denoted by

$$(1.1) \qquad A x = b \qquad\qquad (A: \text{matrix}, \ x, b: \text{vectors}).$$

The equation is connected with a discretisation parameter («grid size»). The discretisation parameter h determines the dimension n_ℓ of the system (1.1). Consider a whole sequence of step sizes:

$$h_0 > h_1 > \ldots > h_{\ell-1} > h_\ell > \ldots \qquad \text{with } \lim h_\ell = 0.$$

Here, ℓ is called the level (number). Often, one can choose $h_\ell := h_0 / 2^\ell$. The discretisation corresponding to level ℓ is denoted by

$$(1.2) \qquad A_\ell x_\ell = b_\ell.$$

72

A_ℓ, x_ℓ, and b_ℓ are matrices and vectors of size n_ℓ. The set of all systems (1.2) for $\ell=0,1,\ldots$ is the hierarchy of discrete problems. The original system (1.1) coincides with (1.2) for a special index ℓ. Further, we need a _prolongation_ (coarse-to-fine mapping) p and a _restriction_ (fine-to-coarse mapping) r.

The multi-grid method is the product of two different iterations: the _smoothing iteration_ $\mathcal{S}_\ell$ (cf. (1.3c)) and the _coarse-grid correction_ (1.3d-f).

(1.3)	**Multi-Grid Procedure** MGM for solving $A_\ell x_\ell = b_\ell$, $\ell \geqslant 0$
(1.3a)	<u>procedure</u> $MGM(\ell, x, b)$; <u>integer</u> ℓ; <u>array</u> x,b;
(1.3b)	<u>if</u> $\ell=0$ <u>then</u> $x := A_0^{-1} b$ <u>else</u>
	<u>begin</u> <u>array</u> d, y; <u>integer</u> i;
(1.3c)	<u>for</u> $i:=1$ <u>to</u> ν <u>do</u> $x := \mathcal{S}_\ell(x, b)$ (smoothing step)
(1.3d)	$d := r*(A_\ell x - b)$; (defect computation)
(1.3e$_1$)	$y := 0$; (set starting value)
(1.3e$_2$)	<u>for</u> $i:=1$ <u>to</u> γ <u>do</u> $MGM(\ell-1, d, y)$; (γ multi-grid calls)
(1.3f)	$x := x - p*y$ (coarse-grid correction)
	<u>end</u>;

Concerning details about the multi-grid method, we refer to [Ha85]. Applications to BEM are also described in [Ha89].

1.2. Application to an Integral Equation of the Second Kind

We consider the Fredholm integral equation

$$\lambda f = Kf + g \quad \text{with} \quad (Kf)(x) = \int_\Gamma k(x,y) f(y)\, d\Gamma_y.$$

The subspace of ansatz functions is denoted by $X_\ell \subset C(\Gamma)$ $(\ell=0,1,2,\ldots)$. Its dimension is $n_\ell = \dim(X_\ell)$. Let $\{\xi_{1,\ell}, \xi_{2,\ell}, \ldots, \xi_{n_\ell,\ell}\}$ be the set of collocation points. Then the collocation solution $f_\ell \in X_\ell$ is defined by

$$\lambda f_\ell(\xi_{j,\ell}) = (Kf_\ell)(\xi_{j,\ell}) + g(\xi_{j,\ell}) \qquad \text{for all } 1 \leqslant j \leqslant n_\ell.$$

Choosing a basis $\{\Phi_{1,\ell},\ldots,\Phi_{n_\ell,\ell}\}$ of X_ℓ we obtain a system (1.2): $A_\ell x_\ell = b_\ell$ of n_ℓ equations for the n_ℓ components of $x_\ell \in \mathbb{R}^{n_\ell}$. The matrix is

$$(1.4) \qquad A_\ell = \lambda I_\ell - K_\ell,$$

where $I_{\ell,jk} = \Phi_{k,\ell}(\xi_{j,\ell})$, $K_{\ell,jk} = (K\Phi_{k,\ell})(\xi_{j,\ell})$. Often, I_ℓ is the identity matrix: $I_\ell = I$ (cf. §4 of [Ha89]). The case of the Galerkin discretisation, is similar.

The equation to be solved has the form $\lambda I_\ell x_\ell = K_\ell x_\ell - b_\ell$, where, for the sake of simplicity, we assume $I_\ell = I$. The well-known Picard iteration reads

$$x_\ell^{\mathrm{old}} \mapsto x_\ell^{\mathrm{new}} := (K_\ell x_\ell^{\mathrm{old}} + b_\ell)/\lambda.$$

The multi-grid method of the second kind is the general algorithm (1.3) applied to the system (1.2) with $A_\ell = I - K_\ell$, where the _smoothing iteration is chosen as the Picard iteration._ The analysis shows that (in almost all situations) it makes no sense to perform more than one smoothing step, i.e., the parameter ν in (1.3c) is chosen as $\nu = 1$. The resulting «multi-grid iteration of the second kind» reads as follows.

<table>
<tr><td>(1.5)</td><td colspan="2">Multi-Grid Procedure MGM for solving $\lambda x_\ell = K_\ell x_\ell + b_\ell$, $\ell \geq 0$</td></tr>
<tr><td>(1.5a)</td><td><u>procedure</u> MGM(ℓ, x, b); <u>integer</u> ℓ; <u>array</u> x, b;</td><td></td></tr>
<tr><td>(1.5b)</td><td><u>if</u> $\ell = 0$ <u>then</u> $x := (\lambda I - K_0)^{-1} b$ <u>else</u></td><td></td></tr>
<tr><td></td><td><u>begin</u> <u>array</u> d, y; <u>integer</u> i;</td><td></td></tr>
<tr><td>(1.5c)</td><td>$x := \frac{1}{\lambda}(b + K_\ell * x)$;</td><td>(Picard iteration)</td></tr>
<tr><td>(1.5d)</td><td>$d := r * (\lambda x - b - K_\ell * x)$;</td><td>(defect computation)</td></tr>
<tr><td>(1.5e$_1$)</td><td>$y := 0$;</td><td>(set starting value)</td></tr>
<tr><td>(1.5e$_2$)</td><td><u>for</u> $i := 1$ <u>to</u> 2 <u>do</u> MGM$(\ell-1, d, y)$;</td><td>(2 multi-grid calls)</td></tr>
<tr><td>(1.5f)</td><td>$x := x - p * y$</td><td>(coarse-grid correction)</td></tr>
<tr><td></td><td><u>end</u>;</td><td></td></tr>
</table>

The general result of the convergence analysis (e.g., presented in [Ha85,§16] and [Ha89,§5]) is

$$\text{multi-grid convergence speed} = O(h^\varkappa)$$

with a positive exponent $\varkappa$, i.e., the convergence speed improves with increasing dimension. For further references, we mention [HeS], [Sch], and [AG].

1.3 Application to Integral Equations of Positive Order

Hypersingular integral equations with kernels $k(x,y) = O(|x-y|^{1-d-\alpha})$ $(\alpha > 0)$ for $\Gamma = \partial\Omega$, $\Omega \subset \mathbb{R}^d$, have the positive order α. Therefore, the integral operators K behave like differential operators. For example, the Poisson problem leads to the hypersingular integral equation

$$(1.6) \qquad \int_\Gamma f(y) \frac{\partial}{\partial n_x} \frac{\partial}{\partial n_y} s(x,y) d\Gamma_y = \varphi(x) \text{ for } x \in \Gamma$$

with the singularity function $s(x,y) = 1/[4\pi|x-y|]$ (cf. [Ha89,§8]). For the theoretical and the numerical treatment it is convenient to replace the Hadamard integral (1.6) by the variational formulation

$$a(f,g) = \int_\Gamma \varphi g \, d\Gamma_y \qquad \text{for all } g \in X,$$

where the bilinear form $a : X \times X \to \mathbb{R}$ is defined by

$$(1.7) \qquad a(f,g) = \frac{1}{2} \iint_{\Gamma \times \Gamma} [g(\mathbf{x}) - g(\mathbf{y})][f(\mathbf{x}) - f(\mathbf{y})] \frac{\partial^2 s(\mathbf{x},\mathbf{y})}{\partial n_x \partial n_y} d\Gamma_y \, d\Gamma_x.$$

Here, a is symmetric and X-elliptic with respect to the Sobolev space $X = H^{1/2}(\Gamma)/\text{const}$: $a(f,f) \geqslant c\|f\|_{1/2}^2$ with some $c>0$ for all $f \in X$.

For elliptic problems of positive order (like (1.7)) the standard multi-grid method applies. The concrete description of the multi-grid method (1.3) requires the following details: (i) Smoothing iteration $\mathcal{S}_\ell$: The simplest choice is the Richardson iteration

$$x_\ell \mapsto \mathcal{S}_\ell(x_\ell, b_\ell) := x_\ell - \frac{1}{\|A_\ell\|}(A_\ell x_\ell - b_\ell).$$

(ii) The number ν of smoothing steps should be sufficiently large as can be seen from the convergence result (1.12). (iii) For γ one may choose 1 as well as 2. As proved in [Ha85,§§6-7] the multi-grid iteration has a convergence speed independent of the step size h_ℓ:

$$(1.8) \qquad \text{multi-grid convergence speed} \leqslant \rho(\nu) := C/\nu^\lambda \qquad (\lambda > 0).$$

The exponent λ is connected with the *regularity* of the operator A associated with the bilinear form a. E.g. for $\lambda = 1$, α-regularity is required: A^{-1}: $L^2(\Gamma) \to H^\alpha(\Gamma)$. For problem (1.7) and Lipschitz domains, the regularity is proved by [Co]. Hence, the estimate (1.8) follows with $\lambda = 1$. Since the form (1.7) is symmetric and elliptic, even the stronger estimate by $\rho'(\nu) < 1$ holds for the V-cycle as well as for the W-cycle (cf. [Ha85, Theorems 7.2.2-3]).

1.4 Transformation into a Problem of Second Kind

Let $A_\ell x_\ell = b_\ell$ be a discrete integral equation of the first kind. Sometimes, A_ℓ can be split into the sum $A_\ell = A_\ell^0 + B_\ell$, where A_ℓ^0 is numerically invertible, while B_ℓ is the discretisation of a smooth integral operator. Then the system $A_\ell x_\ell = b_\ell$ is equivalent to

$$(1.13) \qquad x_\ell = C_\ell x_\ell - c_\ell \quad \text{with } C_\ell := -(A_\ell^0)^{-1}B_\ell, \; c_\ell := -(A_\ell^0)^{-1}b_\ell.$$

This system is formally a problem of the second kind. The multi-grid procedure (1.5) can be applied without any change to problem (1.13). For a simple example, compare [Ha85,§16.10.5] or [Ha91b].

2. The Panel Clustering Method

The work for the matrix-vector multiplication is the measure for the overall work needed for the multi-grid method. The standard implementation requires $O(n^2)$ operations. Another difficulty arising in the definition phase of the BEM is the need of storage of the size $O(n^2)$ for the matrix entries. Here, we give the description of the panel clustering method by which the multiplication can performed in $O(n\log^{d+1}n)$ operations. Also the storage requirements can be reduced to $O(n\log^{d+1}n)$. d is the spatial dimension of the domain in the boundary value problem, i.e., $d=2$ or 3. For details of the panel clustering method we refer to Hackbusch – Nowak [HN86, HN89] and

[Ha90].

In the following, we replace the level index ℓ (e.g. in A_ℓ) by n corresponding to n_ℓ for a fixed ℓ. Let K_n be the matrix in (1.4): $A_n = \lambda I - K_n$ and consider the matrix-vector product. The ith component of the product of K_n with the vector of the nodal values α_j equals the following integral (evaluated at the collocation point ξ^i):

$$\sum_j \alpha_j \int_{\Gamma_n} k(\xi^i, \mathbf{y}) b_j(\mathbf{y}) d\Gamma_\mathbf{y}.$$

Γ_n is the union of so-called *panels* or *boundary elements*. Assembling several neighbouring boundary elements (panels), we can form first clusters. The union of neighbouring clusters yield even larger clusters. This process can be continued until the complete surface Γ_n is obtained as the largest cluster (cf. Fig. 1).

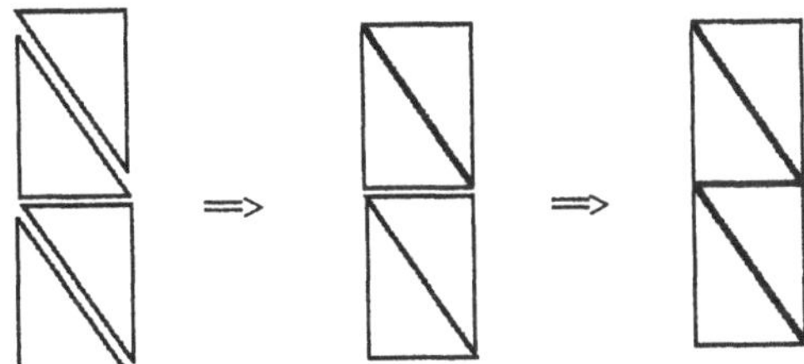

FIG. 1. Clustering of four triangles

For fixed ξ^i, one determines a decomposition of Γ_n into disjoint clusters τ_ℓ, $\ell = 1, ..., \sigma_i$. The size of the clusters τ_ℓ should be chosen proportional to the distance from ξ^i. Under natural assumptions, one obtains the estimate $\sigma_i = O(\log n)$ for the number of clusters. Let $\mathfrak{C}_{\text{near}}(\xi^i) := \mathfrak{C}_{\text{near}}$ be the set of those τ_ℓ, which are panels. The remaining (nontrivial) clusters form the set $\mathfrak{C}_{\text{far}}(\xi^i) = \mathfrak{C}_{\text{far}}$ (for details of the construction compare, e.g., [Ha90]). The integral over Γ_n splits into the sums

$$\int_{\Gamma_n} ... = \sum_{\tau \in \mathfrak{C}_{\text{near}}} \int_\tau ... + \sum_{\tau \in \mathfrak{C}_{\text{far}}} \int_\tau ... \ .$$

By definition, $\tau \in \mathfrak{C}_{\text{near}}(\xi^i)$ is a panel and the integral can be determined as usual. For $\tau \in \mathfrak{C}_{\text{far}}(\xi^i)$, we have to evaluate each term in

$$(2.1) \qquad \sum_{\tau \in \mathfrak{C}_{\text{far}}} \int_\tau k(\xi^i, y) b_j(y) d\Gamma_y.$$

Let the kernel $k(x, y)$ be approximated by a finite expansion (e.g. Taylor's expansion) of the form

$$\tilde{k}(x, y) = \sum_\iota \varkappa_\iota(x; y_0) \Psi_\iota(y),$$

where $\Psi_\iota(y)$ are functions independent of the other quantities x and y_0. The point x will be replaced, e.g., by a collocation point ξ^j. The quantity y_0 denotes the centre of the expansion and may vary in the applications, since $k(x,y)$ is considered over different subsets of the surface Γ_n. Replacing k in (2.1) by $\tilde{k}$, we obtain

$$\sum_{\tau \in \mathfrak{C}_{far}} \int_\tau \tilde{k}(\xi^i, y)\, b_j(y)\, d\Gamma_y =$$
$$= \sum_\tau \int_\tau \sum_\iota \varkappa_\iota(\xi^i; z_\tau)\, \Psi_\iota(y)\, b_j(y)\, d\Gamma_y$$

Summation and integration can be interchanged:

$$\sum_\tau \sum_\iota \varkappa_\iota(\xi^i; z_\tau) \int_\tau \Psi_\iota(y)\, b_j(y)\, d\Gamma_y .$$

The integrals $J_\tau^\iota(b_j) := \int_\tau \Psi_\iota(y) b_j(y) d\Gamma_y$ will be called _far field coefficients_. They can easily be computed from

$$J_\pi^\iota(b_j) := \int_\pi \Psi_\iota(y)\, \Phi_j(y)\, d\Gamma_y$$

Evaluating these far field coefficients in a first phase, we can obtain the desired expression (2.1) by summing over all ι and τ. Since there are only logarithmically many terms in these sums, the required storage and the computational work turns out to be proportional to $n \log^{d+1} n$.

3. Cubature for Galerkin's Method

Collocation leads to the evaluation of integrals of the form $\int k(x,y) b_j(y) d\Gamma_y$. Differently, Galerkin's approach involves the two-fold integral $\iint k(x,y) b_j(y) b_j(x) d\Gamma_y d\Gamma_x$. The advantages of the Galerkin method are its stability and better error estimates with respect to weaker norms. A particular advantage arises for hypersingular integrals, where the singularity of k is of order $|x-y|^{-3}$ (cf. (1.6)). In this case $\int k(x,y) b_j(y) d\Gamma_y$ has to be interpreted in the sense of Hadamard, while $\iint k(x,y) b_j(y) b_j(x) d\Gamma_y d\Gamma_x$ exists as (improper) integral. However, as soon as the x-integral is replaced by some quadrature method, the y-integral is again hypersingular. This fact makes it difficult to apply tensorial methods.

A direct cubature approach to the double integral is described by Sauter [Sa]. He replaces the variables x, y by x and $u = x - y$. While $\iint ... d\Gamma_y d\Gamma_x$ is taken over the product of two triangles, the domain of the transformed double integral turns out to be the sum of six simple regions. The integration with respect to $u = x - y$ can be performed explicitly, whereas the remaining integrands are smooth so that standard Gauß formulae can be used. For details compare [Sa] and [HaS].

4. Cubature for Nearly Singular Integrals

We consider integrals fo the form

$$(4.1) \qquad \int_\Delta f(x_0, x-x_0) \, \|x-x_0\|^{-s-t} d\Gamma_x$$

where $x_0 \epsilon \mathbb{R}^3$ is the source point, Δ a (plane) triangle in $\mathbb{R}^3$ and f with $f(\circ, w) = O(|w|^t)$ a smooth function. Thus, the singularity is of order s. In the case that the singular point is a vertex of the triangle Δ, cubature methods are known. Here, we study the case of x_0 being close to Δ. Then, the techniques mentioned above do not apply, while standard methods are inefficient because of the high (almost infinite) derivatives of the integrand.

Replace $f(x_0, x-x_0)$ by a polynomial $\sum c_\nu(x_0) w^\nu$ in $w = x - x_0$ and transform Δ into a new triangle (named again Δ) lying in the (u_1, u_2)-plane so that $(0,0)$ is one of the vertices and x_0 is mapped into $(-B, 0, H)$. Then, the integral becomes a sum of terms of the form

$$(4.2) \qquad \int_\Delta (u_1+B)^{\nu_1} u_2^{\nu_2} H^{\nu_3} / [(u_1+B)^2 + u_2^2 + H^2]^{(s+t)/2} \, d\Gamma_u \; .$$

Using polar coordinates for u, we obtain an inner integral with respect to the radial coordinate which can be evaluated explicitly. For its representation and a stability discussion compare [HaS93]. The remaining angular integral has a smooth integrand. [HaS93] discusses the size of the remainder of the Gauß quadrature and its (in)dependence of the distance to the singularity.

References.

[AG] ATKINSON, K. and I. G. GRAHAM: Iterative solution of linear systems arising from the boundary integral method. Manuscript

[Co] COSTABEL, M.: Boundary integral operators on Lipschitz domains: elementary results. *SIAM J. Math. Anal. 19* (1988) 613-626

[Ha85] HACKBUSCH, W. *Multi-Grid Methods and Applications.* Springer Heidelberg, 1985

[Ha86] HACKBUSCH, W.: *Theorie und Numerik elliptischer Differentialgleichungen.* Teubner, Stuttgart 1986 – Engl. translation: *Elliptic differential equations.* Springer, Berlin 1992

[Ha89] HACKBUSCH, W.: *Integralgleichungen. Theorie und Numerik.* Teubner, Stuttgart 1989. Engl. translation: *Integral equations.* To be published by Birkhäuser, Basel

[Ha90] HACKBUSCH, W.: The panel clustering algorithm. In: Whiteman, J. R. (ed.): The mathematics of finite elements and applications VII – MAFELAP 1990. Proceedings, Uxbridge, April 1990. Academic Press, London 1990

[Ha91a] HACKBUSCH, W.: *Iterative Lösung großer schwachbesetzter Gleichungssysteme.* Teubner, Stuttgart 1991. - Engl. translation: *Iterative solution of large sparse systems of equations.* Springer, New York 1993

[Ha91b] HACKBUSCH, W.: The solution of large systems of BEM equations by the multi-grid and panel clustering technique. Proceedings, Juni 1990, Turin. *Rend. Sem. Mat. Univ. Pol. Torino, Fasc. Speciale 1991, Numerical Methods.* Pages 163-187.

[HN86] HACKBUSCH, W. and Z.P. NOWAK: O cložnosti metoda panelej (russ.). In: Marčuk, G.I. (ed.): *Vyčislitel'nye prozessy i sistemy.* Proceedings, Moscow, Sept. 1986. Nauka, Moscow 1988. pp. 233-244

[HN89] HACKBUSCH, W. and NOWAK, Z.P.: On the fast matrix multiplication in the boundary element method by panel clustering. *Numerische Mathematik 54* (1989) 463-491

[HaS93] HACKBUSCH, W. and S. SAUTER: On the efficient use of the Galerkin method to solve Fredholm integral equations. *Applications of Mathematics 38* (1993) 301-322

[HaS94] HACKBUSCH, W. and S. SAUTER: On numerical cubature of nearly singular surface integrals arising in BEM collocation. *Computing,* to appear in 1994 (Bericht Nr 93-4, Universität Kiel)

[HeS] HEMKER, P. W. and H. SCHIPPERS: Multiple grid methods for the solution of Fredholm integral equations of the second kind. *Math. Comp. 36* (1981) 215-232

[HKW] HSIAO, G. C., P. KOPP and W. L. WENDLAND: Some applications of a Galerkin-collocation method for boundary integral equations of the first kind. *Math. Meth. in the Appl. Sci. 6* (1984) 280-325

[R] ROKHLIN, V.: Rapid solution of integral equations of classical potential theory. *J. Comput. Physics 60* (1985) 187-207

[Sa] SAUTER, S: Über die effiziente Verwendung des Galerkinverfahrens zur Lösung Fredholmscher Integralgleichungen. Dissertation, Universität Kiel, 1992

[Sch] SCHIPPERS, H.: Multigrid methods for boundary integral equations. *Numer. Math. 46* (1985) 351-363

[Schw] SCHWAB, C: Variable order composite quadrature of singular and nearly singular integrals. Manuscript 1993

Wolfgang Hackbusch
Institut für Informatik und Praktische Mathematik
Christian-Albrechts-Universität zu Kiel
D-24098 Kiel, Germany

On the Boundary-Field Equation Methods for Fluid-Structure Interactions

George C. Hsiao

1 Introduction

This paper is concerned with some recent developments in the combined boundary integral and field equation methods for fluid-structure interaction problems. Emphasis will be placed upon the variational formulations and mathematical foundations of various solution procedures.

In the simplest physical situation, the fluid-structure interaction problem can be described as follows: An acoustic wave propagates in a fluid domain of infinite extent where a bounded elastic structure is immersed. The problem is to determine the scattered acoustic pressure in the fluid domain and the displacement fields in the elastic structure. In the following, let Ω be a bounded domain in $I\!\!R^3$ occupied by the elastic structure. We denote by Γ the boundary of Ω and by $\Omega^c := I\!\!R^3 \backslash \Omega \cup \Gamma$ the exterior unbounded domain filled with a compressible inviscid fluid. For the linear model, by assuming that the wave motion is time-harmonic, the physical problem can then be completely formulated in terms of the displacement field $\mathbf{u}(x)$ and the pressure field $p(x)$ for the elastic structure and the fluid, respectively. The corresponding boundary-value problem consists of the governing equations:

$$\mu \Delta \mathbf{u} + (\lambda + \mu) grad\ div\ \mathbf{u} + \rho \omega^2 \mathbf{u} = \mathbf{0} \quad \text{in} \quad \Omega$$

(E)

$$\Delta p + k^2 p = 0 \quad \text{in} \quad \Omega^c$$

together with the transmission conditions:

$$\mathbf{t}^- = -(p^+ + p^{inc})\mathbf{n}$$

(B)
$$\mathbf{u}^- \cdot \mathbf{n} = \frac{1}{\rho_f \omega^2}\left(\frac{\partial p^+}{\partial n} + \frac{\partial p^{inc}}{\partial n}\right) \quad \text{on} \quad \Gamma$$

and the Sommerfeld radiation condition:

(C)
$$\frac{\partial p}{\partial r} - ikp = o(r^{-1}) \quad \text{as} \quad r = |x| \longrightarrow \infty.$$

80

Here the physical parameters: μ and λ are Lamé constants for the elastic material such that $\mu > 0$ and $3\lambda + 2\mu > 0$; ρ and ρ_f are the densities of the elastic structure and the fluid; ω the frequency, k the wave number and is defined by $k^2 := \omega^2/c^2$ with the speed of sound c in the fluid. As usual, we assume that $Im\ k \geq 0$ and $Re\ k > 0$, if $Im\ k = 0$ so that the Rellich lemma holds for the uniqueness proof. The superscripts $\pm$ notation in (B) standards for taking the limits on Γ from Ω^c and Ω, respectively. The traction $\mathbf{t}^-$ is defined by $\mathbf{t}^- := T[\mathbf{u}]|_\Gamma$ where T is the traction operator

$$T[\mathbf{u}]|_\Gamma := 2\mu\frac{\partial \mathbf{u}}{\partial n} + \lambda\mathbf{n}\nabla\cdot\mathbf{u} + \mu\mathbf{n} \times curl\ \mathbf{u}\Big|_\Gamma.$$

Here and through out the paper $\mathbf{n}$ is the outward unit normal with respect to the bounded domain Ω. The transmission conditions (B) simply state that the limits of the the tractions as well as the normal components of the accelerations from both the elastic structure and the fluid are equal up to the given incident fields p^{inc} and $\partial p^{inc}/\partial n$ (see, e.g [17],[13]).

The boundary-value problem (E), (B), (C) is a transmission problem involving two media, one bounded and one of infinite extent. Problems of this kind are best treated by the methods of the boundary-field equation.

2　Nonlocal Boundary Problems

The basic idea of the boundary-field equation methods is the reduction of the original boundary-value problem for (E) in the exterior domain Ω^c to a boundary integral equation on Γ (or on any auxiliary boundary Γ_0 containing Ω completely in its interior). In this way, the original boundary-value problem is transformed to an equivalent problem, the so-called *non-local boundary problem*, in a bounded region for the partial differential equation (E) (or equations (E)) together with the boundary integral equation on Γ (or on Γ_0). For this equivalent problem, existence, uniqueness as well as approximate solutions can then be established by the conventional variational methods for boundary-value problems in bounded domain.

It is well known that the reduction of boundary-value problems to boundary integral equations is generally not a unique process. Consequently, there are various forms or versions of the non-local boundary problems depending on the derivations of the boundary integral equations. In the following, we shall confine to the non-local boundary problems in the bounded domain Ω without introducing an auxiliary boundary Γ_0. The latter case can be easily modified.

We note that the boundary-value problem for the linear elasticity equation (E) in Ω will be completely decoupled, if one knows the trace of p on Γ (i.e. p^+). Hence it is nature that one should seek an additional equation for the unknown p^+, but not for $\partial p^+/\partial n$ on Γ in contrast to the standard transmission problem (see e.g. [7]).

To illustrate the idea, let us begin with the Green representation formula for p in Ω^c:

$$p(x) \;=\; \int_\Gamma \frac{\partial}{\partial n_y}\gamma(x,y)p^+(y)ds_y - \int_\Gamma \gamma(x,y)\frac{\partial p^+(y)}{\partial n_y}ds_y \tag{1}$$

$$=: \mathcal{D}p^+(x) - \mathcal{S}\frac{\partial p^+}{\partial n}(x), \quad x \in \Omega^c.$$

Here $\gamma(x,y) = e^{ik|x-y|}/4\pi|x-y|$ is the fundamental solution of the Helmholtz equation. By taking the limits of (1) and its normal derivative, as x tends to Γ, and using the well known jump relations for simple and double layer potentials, $\mathcal{S}p^+$ and $\mathcal{D}(\partial p^+/\partial n)$, we arrive at two boundary integral equations:

$$(\frac{1}{2}I - K)p^+ + V(\frac{\partial p^+}{\partial n}) = 0$$

$$\qquad\qquad \text{on} \quad \Gamma. \tag{2}$$

$$Wp^+ + (\frac{1}{2}I + K')(\frac{\partial p^+}{\partial n}) = 0$$

Here V, K, K' and W are the four basic boundary integral operators defined by

$$V\sigma(x) := \int_\Gamma \gamma(x,y)\sigma(y)ds_y, \quad K\mu(x) := \int_\Gamma \frac{\partial}{\partial n_y}\gamma(x,y)\mu(y)ds_y,$$

$$K'\sigma(x) := \int_\Gamma \frac{\partial}{\partial n_x}\gamma(x,y)\sigma(y)ds_y, \quad W\mu(x) := -\frac{\partial}{\partial n_x}\int_\Gamma \frac{\partial}{\partial n_y}\gamma(x,y)\mu(y)ds_y$$

for sufficiently smooth functions σ and μ on Γ. In view of the transmission conditions (B), we may replace $\partial p^+/\partial n$ and rewrite (2) in the form:

$$(\frac{1}{2}I - K)p^+ + \rho_f\omega^2 V(\mathbf{u}^- \cdot \mathbf{n}) = V(\frac{\partial p^{inc}}{\partial n})$$

$$\qquad\qquad \text{on} \quad \Gamma. \tag{3}$$

$$Wp^+ + \rho_f\omega^2(\frac{1}{2}I + K')(\mathbf{u}^- \cdot \mathbf{n}) = (\frac{1}{2}I + K')\frac{\partial p^{inc}}{\partial n}$$

For given $\rho_f\omega^2(\mathbf{u}^- \cdot \mathbf{n})$ and $\partial p^{inc}/\partial n$, these are respectively the boundary integral equations of the second and first kind for the known p^+. The usual approach in the method of combined boundary integral and field equations is to formulate the non-local boundary problem by using the first kind boundary integral equation [15], [21]. However, as will be seen, one may also formulate the problem by taking the second kind boundary integral equation but with appropriate weak formulations. This will become transparent, when we discuss their weak formulations.

We now formulate the two versions of the non-local boundary problem for (E), (B), (C) as follows:

Nonlocal Boundary Problem: *Find $(\mathbf{u}, p^+)$ satisfying*

$$\Delta^* \mathbf{u} + \rho \omega^2 \mathbf{u} = \mathbf{0} \quad \text{in} \quad \Omega, \quad \mathbf{t}^- = -(p^+ + p^{inc})\mathbf{n} \quad \text{on} \quad \Gamma \tag{4}$$

and the non-local boundary condition (NLBC) *on* Γ:

$$W p^+ + \rho_f \omega^2 (\frac{1}{2}I + K')(\mathbf{u}^- \cdot \mathbf{n}) = (\frac{1}{2}I + K')\frac{\partial p^{inc}}{\partial n} \tag{5}$$

or

$$(\frac{1}{2}I - K)p^+ + \rho_f \omega^2 V(\mathbf{u}^- \cdot \mathbf{n}) = V(\frac{\partial p^{inc}}{\partial n}) \quad . \tag{6}$$

In the formulation, we have adopted the notation $\Delta^* \mathbf{u}$ for the Lamé operator:

$$\Delta^* \mathbf{u} := \mu \Delta \mathbf{u} + (\lambda + \mu) grad\ div\ \mathbf{u}.$$

We remark that condition (5) or (6) is usually referred as a *non-local boundary condition*, since the values $\mathbf{u}^- \cdot \mathbf{n}$ over the entire boundary Γ are needed in order to compute p^+ at a single point $x \in \Gamma$ (see e.g. [19], [11]).

3 Variational Formulations

As aforementioned, the difference between the two versions of the non-local boundary problem (4),(5) and (4), (6) lies in their weak formulations. The usual approach [3] based on the L^2-weak formulation of the second kind equation without any modification is no longer an appropriate one. In particular, we note that the proper solution space for p^+ is not $L^2(\Gamma)$ but the Sobolev space $H^{1/2}(\Gamma)$.

The variational formulation of the non-local boundary problem (4),(5),(6) reads:
Given $p^{inc} \in H^{1/2}(\Gamma)$, find $(\mathbf{u}, p^+) \in (H^1(\Omega))^3 \times H^{1/2}(\Gamma)$ such that

$$a(\mathbf{u}, \mathbf{v}) - \rho \omega^2 (\mathbf{u}, \bar{\mathbf{v}})_0 + <p^+, \mathbf{n} \cdot \bar{\mathbf{v}}^- >_0 = - <p^{inc}, \mathbf{n} \cdot \bar{\mathbf{v}}^- >_0 \tag{7}$$

and either

$$< Wp^+, \bar{q} >_0 + \rho_f \omega^2 < (\frac{1}{2}I + K')(\mathbf{u}^- \cdot \mathbf{n}), \bar{q} >_0$$
$$=< (\frac{1}{2}I + K')\frac{\partial p^{inc}}{\partial n}, \bar{q} >_0 \tag{8}$$

or

$$< (\frac{1}{2}I - K)p^+, \bar{q} >_{1/2} + \rho_f \omega^2 < V(\mathbf{u}^- \cdot \mathbf{n}), \bar{q} >_{1/2}$$
$$=< V(\frac{\partial p^{inc}}{\partial n}), \bar{q} >_{1/2} \tag{9}$$

for all $(\mathbf{v}, q) \in (H^1(\Omega))^3 \times H^{1/2}(\Gamma)$.

Here $a(\cdot \cdot)$ is the sesquilinear form corresponding to the Lamé operator $-\Delta^*$,

$$a(\mathbf{u}, \mathbf{v}) := \int_\Omega \left\{ \frac{3\lambda + 2\mu}{3} div\ \mathbf{u}\ \overline{div\ \mathbf{v}} + \frac{\mu}{2} \sum_{i \neq j}^3 (\frac{\partial u_i}{\partial x_j} + \frac{\partial u_j}{\partial x_i})(\overline{\frac{\partial v_i}{\partial x_j} + \frac{\partial v_j}{\partial x_i}}) \right.$$
$$\left. + \frac{\mu}{3} \sum_{i,j=1}^3 (\frac{\partial u_i}{\partial x_j} - \frac{\partial u_j}{\partial x_i})(\overline{\frac{\partial v_i}{\partial x_j} - \frac{\partial v_j}{\partial x_i}}) \right\} dx,$$

while $(\cdot, \cdot)_0$ denotes the L^2-duality pairing on $(H^1(\Omega))^3$ and its dual, $< \cdot, \cdot >_0$ the L^2- duality pairing on $H^{1/2}(\Gamma) \times H^{-1/2}(\Gamma)$, and $< \cdot, \cdot >_{1/2}$, the $H^{1/2}(\Gamma)$-inner product on $H^{1/2}(\Gamma)$.

Following [10], we have the following result.

Theorem 1 *Let* $A((\mathbf{u}, p^+), (\mathbf{v}, q))$ *be the sesquilinear form defined by either* (7), (8) *or* (7), (9). *In either case, there holds the Gårding inequality:*

$$Re\ A((\mathbf{u}, p^+), (\mathbf{u}, p^+)) \quad \geq \quad \alpha \left\{ \|\mathbf{u}\|^2_{(H^1(\Omega))^3} + \|p^+\|^2_{H^{1/2}(\Gamma)} \right\}$$
$$- \quad Re\ C((\mathbf{u}, p^+), (\mathbf{u}, p^+))$$

for all $(\mathbf{u}, p^+) \in (H^1(\Omega))^3 \times H^{1/2}(\Gamma)$, *where* C *is a compact sesquilinear form on* $(H^1(\Omega))^3 \times H^{1/2}(\Gamma)$ *and* $\alpha > 0$ *is a constant independent of* $(\mathbf{u}, p^+)$.

Gårding's inequality implies the Fredholm alternative is applicable and existence of a weak solution then follows from uniqueness of the original transmission problem. However, it is known that the original problem is not always uniquely solvable (see e.g. [16], [18]). The existence of *traction free* rotational oscillations for certain values of $\rho\ \omega^2$ and certain geometries is a fundamental difficulty of the formulation of the boundary-value problem (E), (B), (C).

Furthermore, it is well known in the scattering theory that given $\partial p^+/\partial n$ the boundary integral equations (2) for p^+ are not uniquely solvable when k is an *exceptional value*. By this we mean that k is an eigenvalue of the interior Neumann problem for the Helmholtz equation in Ω when we are dealing with first kind boundary integral equations, and k is an eigenvalue of the interior Dirichlet problem in the case of second kind boundary integral equations.

We now summarize these discussions in the following theorem. Details of the proofs will be available in [14].

Theorem 2 *Assume that the homogeneous boundary-value problem of* (E), (B), (C) *has no traction free solution and that k is not an exceptional value. Then the corresponding non-local boundary problem* (4), (5) *or* (4), (6) *has a unique solution* $(\mathbf{u}, p^+) \in ((H^1(\Omega))^3 \times H^{1/2}(\Gamma)$.

Method	Rep. of p in Ω^c	Traction t on Γ
1		$-(p^+ + p^{inc})\mathbf{n}$
2(a)	$p = \mathcal{D}p^+ - \mathcal{S}\sigma$	*(same as above)*
2(b)	$p^+ \in H^{1/2}(\Gamma), \quad \sigma \in H^{-1/2}(\Gamma)$	*(same as above)*
3	$\sigma := \rho_f \omega^2 (\mathbf{u}^- \cdot \mathbf{n}) - \frac{\partial p^{inc}}{\partial n}$	$-((\frac{1}{2}I + K)p^+ - V\sigma + p^{inc})\mathbf{n}$
4		$-(p^+ + p^{inc})\mathbf{n}$
5	$p = \mathcal{D}\mu, \quad \mu \in H^{1/2}(\Gamma)$	$-((\frac{1}{2}I + K)\mu + p^{inc})\mathbf{n}$
6(a)	$p = -\mathcal{S}\phi$	$-(-V\phi + p^{inc})\mathbf{n}$
6(b)	$\phi \in H^{-1/2}(\Gamma)$	*(same as above)*
7	$p = \mathcal{D}\mu + i\eta\mathcal{S}\mu, \quad \mu \in H^{1/2}(\Gamma)$	$-((\frac{1}{2}I + K)\mu + i\eta V\mu + p^{inc})\mathbf{n}$

Method	NLBC on Γ	Weak Form	
1	$Wp^+ + (\frac{1}{2}I + K')\sigma = 0$	$< \cdot, \overline{q} >_0 = 0,$	$q \in H^{1/2}(\Gamma)$
2(a)	$(\frac{1}{2}I - K)p^+ + V\sigma = 0$	$< \cdot, q >_{1/2} = 0,$	$q \in H^{1/2}(\Gamma)$
2(b)	*(same as above)*	$< \cdot, \overline{Wq} >_0 = 0,$	$q \in H^{1/2}(\Gamma)$
3	$Wp^+ + (\frac{1}{2}I + K')\sigma = 0$	$< \cdot, \overline{q} >_0 = 0,$	$q \in H^{1/2}(\Gamma)$
4	$Wp^+ + (\frac{1}{2}I + K')\sigma$ $+i\eta((\frac{1}{2}I - K)p^+ + V\sigma) = 0$	$< \cdot, \overline{q} >_0 = 0,$	$q \in H^{1/2}(\Gamma)$
5	$W\mu + \sigma = 0$	$< \cdot, \overline{\nu} >_0 = 0 ,$	$\nu \in H^{1/2}(\Gamma)$
6(a)	$(\frac{1}{2}I - K')\phi - \sigma = 0$	$< \cdot, \psi >_{-1/2} = 0,$	$\psi \in H^{-1/2}(\Gamma)$
6(b)	*(same as above)*	$< \cdot, \overline{V\psi} >_0 = 0,$	$\psi \in H^{1/2}(\Gamma)$
7	$W\mu + i\eta(\frac{1}{2}I - K')\mu + \sigma = 0$	$< \cdot, \overline{\nu} >_0 = 0,$	$\nu \in H^{1/2}(\Gamma)$

Table 1: Various Non-local Boundary Conditions.

4 Generalizations and Concluding Remarks

Up to now we have derived the non-local boundary conditions (5) and (6) by the *direct method* based on the Green representation formula (1). Clearly one may also obtain non-local boundary conditions by the *indirect method* base on the layer ansatz. In fact, one can combine both the direct and the indirect approaches, as in [2] and [12], to generate a variety of combinations. In all the cases we may obtain similar results as given in Theorems 1 and 2. We remark that the non-uniqueness caused by the exceptional values of the boundary integral equations are not a serious drawback. As in the scattering theory, standard techniques can be employed to eliminate these exceptional values (see e.g. [6], [5] and [20]).

In Table 1 above, we have listed all the available non-local boundary conditions in terms of the boundary integral equations. In particular, we have indicated in the last column of the table how to form properly their weak formulations table. The corresponding traction term t has been also included in the table. This together with the partial differential equation for the elastic structure in the domain Ω will be used to formulate the variational equation. We remark that the boundary integral equations for methods 4 and 7 have no exceptional values and here η is any real constant. These are all first kind boundary integral equations.

To conclude the paper, we comment that one may also introduce an additional auxiliary boundary Γ_0 to enclose the elastic structure. In particular, when Γ_0 is chosen to be a sphere, one may replace the non-local boundary condition by a local one according to the approaches in [8] and [4], and Fourier series development may also be employed [9]. In this connections, we refer to some recent work in [1].

Acknowledgment

This work was supported by the ONR under grant N00014-91-J-1700.

References

[1] Barry, A.; Bielak, J.; MacCamy, R.C.: On absorbing boundary conditions for wave propagation. J. Comput. Phys **79**(1988), 449–468.

[2] Bielak, J.; MacCamy, R.C.: Symmetric finite element and boundary integral coupling methods for fluid-solid interaction. Quart. Appl. Math. **49**(1991), 107–119.

[3] Bielak, J.; MacCamy, R.C.; Zeng, X.: Stable coupling method for interface scattering problems. Research Report No. R-91-199, Dept. of Civil Engineering , Carnegie Mellon University 1991.

[4] Bayliss, A.; Turkel, E.: Radiation boundary conditions for wave-like equations. Comm. Pure Appl. Math. **33**(1980) 6, 707–725.

[5] Brakhage, H.; Werner, P.: Über das Dirichletsche Aussenraumproblem für die Helmholtzsche Schwingungsgleichung. Archiv Mathematik, **16** (1965), 325–329.

[6] Burton, A.J.; Miller, G.F.: The application of integral equation methods to the numerical solutions of some exteror boundary-value problems. Proc. Royal Soc. London Ser. A **323** (1971), 201–210.

[7] Costabel, M.; Stephan, E.P.: Coupling of finite elements and boundary elements for transmission problems of elastic waves. In: Cruse, T. et al. (eds.): Advanced Boundary Element Methods, IUTAM Symposium. San Antonio, Texas 1987, pp. 118–124.

[8] Engquist, B.; Majda, A.: Absorbing boundary conditions for the numerical simulation of waves. Math. Comp. **31**(1977), 629–651.

[9] Feng, K.: Asymptotic radiation conditions for reduced wave equation. J. Comp. Math. **2** (1984) 2, 130–138.

[10] Hsiao, G.C.: The coupling of BEM and FEM - a brief review. In: Brebbia, C.A. et al.(eds): Boundary Elements X, Vol.I. Springer-Verlag, Berlin 1988, pp. 431–445.

[11] Hsiao, G.C.: The coupling of boundary element and finite element methods. Z. angew. Math. Mech. (ZAMM) **70** (1990) 6, T493-T503.

[12] Hsiao, G.C.: Some recent developments on the coupling of finite element and boundary element methods. Rend. Sem. Mat. Univ. Pol. Torino Fascicolo Speciale 1991, Numerical Methods, pp. 95–111.

[13] Hsiao, G.C.; Kleinman, R.E.; Schuetz, L.S.: On variational formulations of boundary value problems for fluid-solid interactions. In: McCarthy, M.F.; Hayes, M.A.(eds.): Elastic Wave Propagation. IUTAM Symposium on Elastic Wave Propagation. North-Holland-Amsterdam 1989, pp. 321–326.

[14] Hsiao, G.C.; Kleinman, R.E.; Roach, G.F.: Weak solutions of fluid-solid interaction problems. In preparation.

[15] Johnson, C.; Nedelec, J.C.: On the coupling of boundary integral and finite element methods. Math. Comp. **35** (1980), 1063–1079

[16] Jones, D.S.: Low-frequency scattering by a body in lubricated contact. Quart. J. Mech. Appl. Math. **36** (1983), 111–137.

[17] Junger, M.C.; Feit, D.: Sound, Structures and Their Interaction. MIT Press, Cambridge, MA 1986.

[18] Kupradze, V.D.: Potential Methods in the Theory of Elasticity. Israel Program for Scientific Translations 1965.

[19] MacCamy, R.C.: Marin, S.P.: A finite element method for exterior interface problems. Int. J. math. and Math. Sci. **3** (1980), 311–350.

[20] Schenck, H.A.: Improved integral formulation for acoustic radiation problems. J. Acoust. Soc. Amer. **44** (1968), 41–58.

[21] Wendland, W.L.: On asymptotic error estimates for the combined BEM and FEM. In: Stein, E.; Wendland, W.L. (eds.): Finite Element and Boundary Element Techniques from Mathematical and Engineering Point of View. CISM Lecture Notes 301. Udine, Springer-Verlag (Wien-New York) 1988, pp. 273–333.

George C. Hsiao
Department of Mathematical Sciences
University of Delaware
Newark, Delaware 19716
U S A
e-mail: hsiao@math.udel.edu

Some Solved and Unsolved Canonical Transmission Problems of Diffraction Theory

Erhard Meister

1 Introduction

The study of the refraction and scattering of waves by penetrable objects has been of large interest to physicists and mathematicians since about one century. Many problems coming from microwave techniques, non-destructive testing theory, geophysics, and other fields of engnineering raised new questions particularly with respect to the asymptotic behavior of scalar and rectorial wave-fields near geometrical singularities, like edges and vertices, on one side, and of the far-field patterns, on the other side. Most recently, the small- and long-time behavior of periodic time-dependent scattered fields is in the center of mathematical interest. In this talk an overview of the state of the art will be given and a number of unsolved problems listed.

This paper is organized as follows. In section two the general transmission problem for the d'Alembert wave equation in n–space is formulated, then the special case of two different media in the upper and lower half-space $\mathbb{R}^n_{\pm}$, respectively, is treated. This section then closes with the *"two-media Sommerfeld half-plane problems"* leading to two-by-two Wiener-Hopf functional systems.

Section three is devoted to wedge-type problems starting with the four quadrant transmission problem for electromagentic waves which are polarized parallel to the common edge. Using a two-dimensional (distributional) Fourier transform for the four scattered wave potentials expressed by their Cauchy boundary data written as one-dimensional unilateral Fourier transforms an equivalent *"four-part Wiener-Hopf functional equation"* may be derived. In a following subsection this will be reduced to a four-by-four system of integral equations for the one-dimensional, unilateral Cosine-Fourier transforms of the normal derivatives. The last subsection will be concerned with some multi-media scattering problems for canonical domains with a Sommerfeld half-plane or right-angled wedge inserted as objects on which boundary conditions have to be fulfilled.

The concluding remarks collect some open problems and challenges for future research.

2 Transmission Problems for the Wave Equation in R^n

2.1 Formulation of the general transmission problem in n-space

Let there be given a time-harmonic primary (incoming) scalar wave-field $Re\,[\Phi_{pr}(\underline{x}; k_1)e^{-i\omega t}]$ in a domain $\Omega_1 \subset \mathbf{R}^n$; $r = 2$ or 3; with boundary $\Gamma = \partial\Omega_1$ and a finite number $\Omega_2, \ldots, \Omega_N$ of disjoint domains such that $\bigcup\limits_{j=1}^{N} \overline{\Omega}_j = \mathbf{R}^n$. Find then the scattered field $Re\,[\Phi_{sc}(\underline{x})e^{i\omega t}]$ in $\mathbf{R}^n$ such that

$$(\Delta_n + k_j^2)\Phi_{sc}(\underline{x})|_{\Omega_j} = 0 \quad \text{for } j = 1, \ldots, N \tag{2.1}$$

with $\Phi_{sc,j}(\underline{x}) \in C^r(\overline{\Omega}_j \backslash E_j)$ or $\Phi_{sc,j} \in H^1_{loc}(\Omega_j)$, the E_j denoting the edges of $\partial\Omega_j$. Across common boundaries $\Gamma_{j\ell} := \partial\Omega_j \cap \partial\Omega_\ell \ (\neq 0)$ we have

$$\Phi_{sc,j}(\underline{x}) - \Phi_{sc,\ell}(\underline{x}) = F_{j\ell}(\underline{x}) \tag{2.2.a}$$

$$\rho_j \frac{\partial \Phi_{sc,j}}{\partial n_j}(\underline{x}) + \rho_\ell \frac{\partial \Phi_{sc,\ell}}{\partial n_\ell} = G_{j\ell}(\underline{x}) \tag{2.2.b}$$

with given data $F_{j\ell}(\underline{x})$ and $G_{j\ell}(\underline{x})$ in correct trace-spaces calculated from the primary wave-field data. For $r_{j\ell} \to 0$ – approaching an edge – the condition of a *"local finite energy"* should hold while for $r \to \infty$ the far-field patterns should behave like

$$\Phi_{sc,j}(\underline{x}) = o(e^{-imk_j \cdot r}) \quad \text{and} \quad \left(\frac{\partial}{\partial r} - ik_j\right)\Phi_{sc,j}(\underline{x}) = o\left(\frac{e^{-imk_j \cdot r}}{\frac{n-1}{2}}\right). \tag{2.3}$$

This general transmission problem has been treated by R. MOHR [14] (1976) in his thesis with smooth interfaces $\Gamma_{j\ell}$ and for Maxwell's equations by C.

WEBER [19] (1977) using Hilbert space methods. Admitting Lipschitz-type boundary (including piece-wise smooth ones), M. COSTABEL and E. STEPHAN [1] reduced the transmission problem in the scalar case (1985) and the electromagnetic case [2] (1988), respectively, to systems of strongly elliptic boundary integral equations for the unknown Cauchy data which are involved in the respresentation formulae for the fields in the domains Ω_j. A generalization to mixed boundary-transmission problems including semi-infinite domains Ω_j and systems of equations having representation formulae – like the Lamé equations of elasto-dynamics – have been studied by T. v. PETERSDORFF [15, 16] (1989a,b), including numerical approximation schemes for the boundary integral equations in his thesis (1989b).

2.2 The simplest transmission problem for the two half-spaces

Let there be given the following geometric situation (see fig. 1) of a plane wave $\Phi_{pr}(\underline{x}; k_1) = \exp\{ik_1(x\cos\theta_1 + y\sin\theta_1)\}$ coming from $\Omega_1 = \mathbf{R}_-^n$ and being partially reflected to $\Phi_{refl}(\underline{x}; k_1) = R\cdot\exp\{ik_1(x\cos\theta_1 - y\sin\theta_1)\}$ in Ω_1 and partially refracted to $\Phi_{trans}(\underline{x}; k_2) = T\cdot\exp\{ik_2(x\cos\theta_2 + y\sin\theta_2)\}$ in $\Omega_2 = \mathbf{R}_+^1$. The transmission conditions (2.2.a), (2.2.b) then lead to

$$k_1\cos\theta_1 = k_2\cos\theta_2 : \quad \text{"Snellius' law"} \tag{2.4}$$

and to the values of the

$$R = \frac{\rho_1\cdot\tan\theta_1 - \rho_2\cdot\tan\theta_2}{\rho_1\cdot\tan\theta_1 + \rho_2\cdot\tan\theta_2} : \quad \text{"reflection coefficient"}, \tag{2.5.a}$$

$$T = \frac{2\rho_1\tan\theta_1}{\rho_1\cdot\tan\theta_1 + \rho_2\cdot\tan\theta_2} : \quad \text{"transmission coefficient"}. \tag{2.5.b}$$

It is well-known that there might exist a total reflection at $\partial\mathbf{R}_\pm^n$ ($\overset{\wedge}{=} x_n = 0$) when $k_2 < k_1$ with a limiting angle $\theta_{1\cdot tot} := \arccos k_2/k_1$ and the transmitted field $\Phi_2(\underline{x}; k_2)$ being exponentially damped in the direction vertical to the interface.

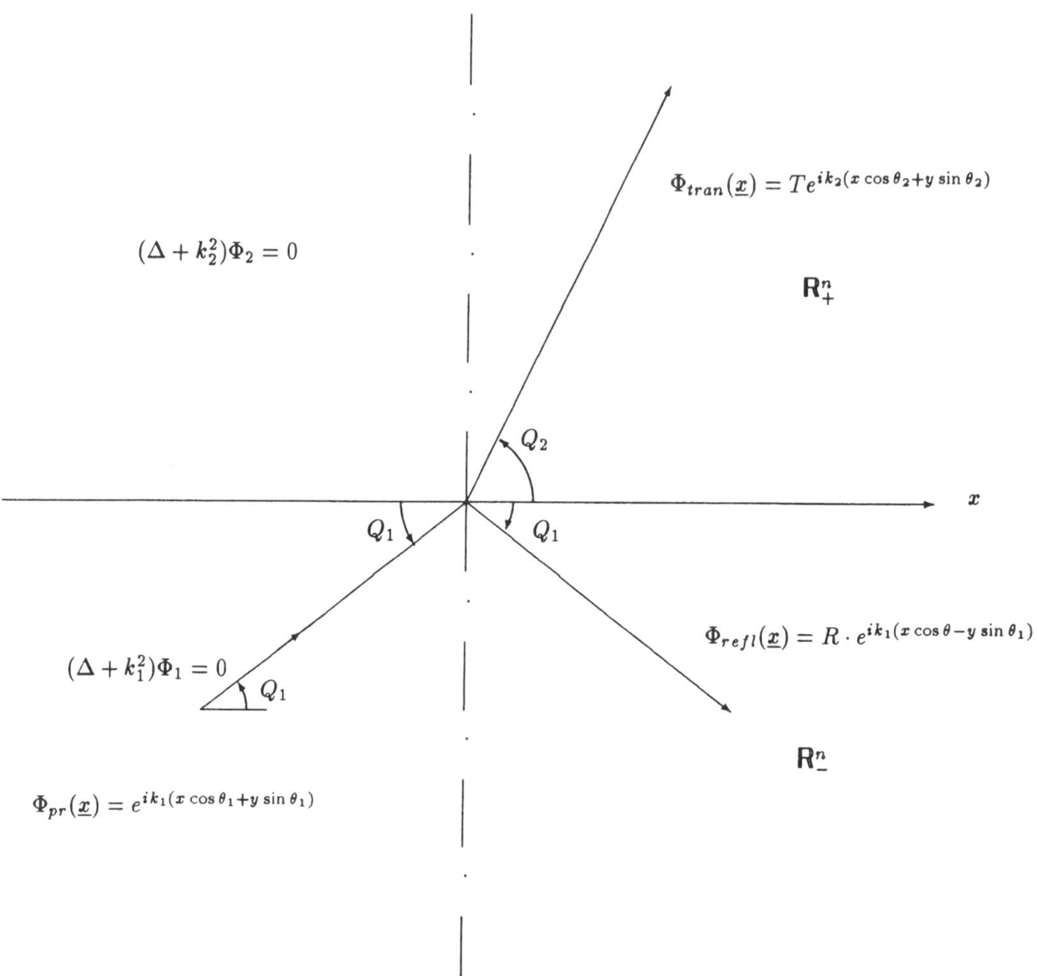

Figure 1: Reflection and refraction of a plane wave
by a plane interface ($\stackrel{\wedge}{=} y = 0$)

2.3   The Sommerfeld half-plane problems for two media

Let there be inserted a half-plane $\mathcal{S} := \{\underline{x} \in \mathbf{R}^3 : x \geq 0\,,\, y = 0\,,\, z \in \mathbf{R}\}$ at the interface $\stackrel{\wedge}{=} y = 0$, of the two media in $\mathbf{R}^3_\pm$. On the two banks $\mathcal{S}_\pm$ of the semi-infinite thin screen $\mathcal{S}$ – due to the polarization type – a Dirichlet or Neumann condition for the scalar wave-field has to be fulfilled, additionally. Applying a one-dimensional (distributional) Fourier transform w.r.t. $x \mapsto \lambda$ (treating here the two-dimensional case, i.e. neglecting the z–dependence!) one arrives to the respresentation formula for the $\mathcal{F}$–transformed fields for $y \lessgtr 0$:

$$\hat{\Phi}_{sc,1/2}(\lambda, y) = A_{1/2}(\lambda) \cdot \exp\left\{ \pm y\sqrt{\lambda^2 - k_{1/2}^2} \right\} \tag{2.6}$$

$$= \begin{cases} (\hat{F}_{1/2+}(\lambda) + \hat{E}_{1/2-}(\lambda)) \cdot \exp\left\{ \pm y\sqrt{\lambda^2 - k_{1/2}^2} \right\} \\ \qquad\qquad\qquad \text{for } D\text{–conditions on } y = 0 \\[2ex] -(\hat{G}_{1/2+}(\lambda) + \hat{V}_{1/2-}(\lambda)) \cdot sgny \cdot \dfrac{\exp\{\pm y\sqrt{\lambda^2 - k_{1/2}^2}\}}{\sqrt{\lambda^2 - k_{1/2}^2}} \\ \qquad\qquad\qquad \text{for } N\text{–conditions on } y = 0 \end{cases}$$

There are known – from the primary wave-field – either

1. $\hat{F}_{1+}$ and $\hat{F}_{2+}$ for the DD–problem on $\mathcal{S}_\pm$

2. $\hat{G}_{1+}$ and $\hat{G}_{2+}$ for the NN–problem on $\mathcal{S}_\pm$

3. $\hat{G}_{1+}$ and $\hat{F}_{2+}$ for the ND–problem on $\mathcal{S}_\pm$

These lead then to – equivalent in the appropriate Sobolev trace spaces $\mathcal{F}$–transformed – "*two-by-two Wiener-Hopf functional equations*" which take the following explicit form in the special case of the DD–problem:

$$\begin{pmatrix} \hat{E}_{1-}(\lambda) \\ \hat{V}_{1-}(\lambda) \end{pmatrix} \begin{pmatrix} \dfrac{\rho_1}{\rho_1\gamma_1 + \rho_2\gamma_2} & , & \dfrac{\rho_2}{\rho_1\gamma_1 + \rho_2\gamma_2} \\[2ex] \dfrac{-\rho_2\gamma_2}{\rho_1\gamma_1 + \rho_2\gamma_2} & , & \dfrac{-\rho_2\gamma_1}{\rho_1\gamma_1 + \rho_2\gamma_2} \end{pmatrix} \begin{pmatrix} \hat{G}_{1+}(\lambda) \\ \hat{G}_{2+}(\lambda) \end{pmatrix} = \begin{pmatrix} \hat{r}_1 \\ \hat{r}_2 \end{pmatrix} \tag{2.7}$$

with

$$\gamma_j := \gamma(\lambda; k_j) := \sqrt{\lambda^2 - k_j^2}, \quad j = 1,2; \tag{2.8}$$

having positive real parts in the complex λ–plane cut from $\lambda = \pm ik_j$ to $\pm i\infty$. The $\hat{r}_{1/2}(\lambda)$ contain given data. The – still unsolved – problem is to factorize the two-by-two "*symbol matrix*" on the right-hand side into non-singular ones which may be continued analytically into the complex half-planes $Im\,\lambda_{\gtrless} \mp Im\,k_j$, respectively, together with their inverses and which behave algebraically for $|\lambda| \to \infty$ there. In the special case of equal media, i.e. $k_1 = k_2$, the factorizations in all three cases of boundary conditions on $\mathcal{S}_\pm$ have been achieved.

Corresponding problems for the Lamé systems of elastodynamics with the same boundary conditions on both sides of the semi-infinite crack $\mathcal{S}$ have been solved

94

by explicit factorization of the three-by-three symbol matrices by the author and F.-O. SPECK [9] (1989). These authors have also studied more general boundary-transmission conditions on $y = 0$ for $\mathcal{S}_\pm$ and its complementary screen $\mathcal{S}'_t$ ($\hat{=} x \leq 0$) and for systems of parallel semi-infinite screens $\mathcal{S}_j$; $j = 1, \ldots, N$; together with K. ROTTBRAND [17] (1991).

3 Transmission problems for right-angled wedges

3.1 The four quadrant transmission problem

Now it is assumed that the whole $\mathbf{R}^3$–space and its cross section $\mathbf{R}^2$ ($\hat{=} z = 0$) is decomposed into the four right-angled wedges Q_j with boundaries $\partial Q_j = \mathcal{S}_j \cup \mathcal{S}_{j+1} \ (mod\,4)$ where homogeneous transmission conditions type (2.2.a) and (2.2.b) are to be fulfilled on the common interfaces for the total wave fields $\Phi_{j,tot}(\underline{x})$, $x \in Q_j$. Applying a two-dimensional (distributional) Fourier transform

$$\hat{\Phi}_j(\lambda_1, \lambda_2) := \iint\limits_{Q_j} e^{i(\lambda_1 x + \lambda_2 y)} \Phi_j(x, y) d(x, y) \tag{3.1}$$

for the scattered wave-fields one may represent those by their one-dimensional unilateral $\mathcal{F}$–transformed Cauchy-boundary data on $\mathcal{S}_1, \ldots, \mathcal{S}_4$, e.g. for

$$\hat{\Phi}_1(\lambda_1, \lambda_2) = [i\lambda_2 \cdot \hat{f}_1^{(1)}(\lambda_1) + \hat{g}_1^{(1)}(\lambda_1) + i\lambda_1 \cdot \hat{f}_2^{(1)}(\lambda_2) + \hat{g}_2^{(1)}(\lambda_2)] \cdot (\lambda_1^2 + \lambda_2^2 + k_1^2)^{-1} \tag{3.2}$$

holding for $0 < (Im\,\lambda_1)^2 + (Im\,\lambda_2)^2 < (Im\,k_1)^2$ and with similar formulae for $\hat{\Phi}_2, \hat{\Phi}_3$ and $\hat{\Phi}_4$ in pairs of complex half-planes $\subset \mathbf{C}^2$.

Due to the transmission conditions holding on $\mathcal{S}_j$; $j = 1, \ldots, 4$; the sum of the four numerators in the corresponding representation formulae (3.2) for $\hat{\Phi}_j(\lambda_1, \lambda_2)$ is known: $\sum_{j=1}^4 Z_j(\lambda_1, \lambda_2) = Z(\lambda_1, \lambda_2)$. Dividing this sum by the "auxiliary characteristic polynomial" $N(\lambda_1, \lambda_2; k) := (\lambda_1^2 + \lambda_2^2 - k^2)$ with an appropriate "artificial wave-number" $k \in \mathbf{C}_+$ one arrives at the "four-part Wiener-Hopf functional equation" in the pair of strips $0 < (Im\,\lambda_1)^2 + (Im\,\lambda_2)^2 < q^2$ where $q := \min_j (Im\,k_j) > 0$.

$$\sum_{j=1}^4 \left(1 + \frac{k^2 - k_j^2}{\lambda_1^2 + \lambda_2^2 - k^2}\right) \hat{P}_j \hat{\Phi}(\lambda_1, \lambda_2) = \frac{Z(\lambda_1, \lambda_2)}{N(\lambda_1, \lambda_2; k)} \tag{3.3}$$

where

$$\hat{\Phi}_j(\lambda_1, \lambda_2) = \hat{P}_j\hat{\Phi}(\lambda_1, \lambda_2) := ((F\chi_{Q'_j}\mathcal{F}^{-1})\hat{\Phi})(\lambda_1, \lambda_2) \tag{3.4}$$

with the characteristic functions χ_{Q_j} of the j-th quadrant Q_j. Acting on the right spaces these continuous projectors $\hat{P}_j$ may be expressed e.g. by

$$\hat{P}_1 = \frac{1}{2}(I + S_1)\frac{1}{2}(I + S_2) \tag{3.5}$$

involving the *"partial Hilbert transforms"* S_1, S_2 w.r.t. the 1st and 2nd variable $\lambda \in \mathbf{R}$. N. LATZ [3] showed (1968) in his thesis that the system (3.3) is uniquely solvable if all $Im\, k_j > 0$ by choosing $k \in \mathbf{C}_+$ ($Re\, k < 0$) appropriately, even in the case of $N \geq 2$ sector-type cross-sections Q_j and he extended his reasoning [4] (1974) to the electro-magnetic case in his habilitation's thesis.

The famous *"dielectric wedge problem"* which results in the case of $k_1 \neq k' = k_2 = k_3 = k_4$ has been studied quite recently by the author, F. PENZEL, F.-O. SPECK and F.S. TEIXEIRA [13], by reduction to a system of integral equations for the unilateral one-dimensional $\mathcal{F}$-transforms of the Cauchy-data on $\mathcal{S}_1$ and $\mathcal{S}_2$.

Here we want to explain a bit the way to get this system.

Start off with formulae (3.2) for $\hat{\Phi}_1$ and $\hat{\Phi}_2$, apply a symmetrization w.r.t. λ_2, what means, take an $\mathcal{F}$-cosine transform w.r.t. λ_2, and then add up to

$$(\lambda_1^2 + \lambda_2^2 - k_1^2)\hat{\Phi}_{1,c}(\lambda_1, \lambda_2) + (\lambda_1^2 + \lambda_2^2 - k_2^2)\hat{\Phi}_{2,c}(\lambda_1, \lambda_2)$$
$$= \hat{g}_1^{(1)}(\lambda_1) + \hat{g}_3^{(2)}(\lambda_1) + \hat{r}_{1,c}(\lambda_1, \lambda_2) \tag{3.6}$$

where now only the $\hat{g}_1^{(1)}$ and $\hat{g}_3^{(2)}$ are unknown. After splitting

$$(\lambda_1^2 + \lambda_2^2 - k_\nu^2) = (\lambda_1 - i\sqrt{\lambda_2^2 - k_\nu^2})(\lambda_1 + i\sqrt{\lambda_2^2 - k_\nu^2}) \tag{3.7}$$

and dividing eq. (3.6) by $(\lambda_1 - i\sqrt{\lambda_2^2 - k_1^2})(\lambda_1 + i\sqrt{\lambda_2^2 - k_2^2})$ and applying then the projectors $\hat{P}_+$ and $\hat{P}_-$ w.r.t. $Im\, \lambda_1 > -q$, $Im\, \lambda_1 < +q$, gives

$$\hat{\Phi}_{1/2,c}(\lambda_1, \lambda_2) = \frac{\lambda_1 \pm i\sqrt{\lambda_2^2 - k_2^2}}{\lambda_1 \pm i\sqrt{\lambda_2^2 - k_1^2}} \cdot \hat{P}_{\pm}\Big([(\sigma - i\sqrt{\lambda_2^2 - k_1^2})(\sigma + i\sqrt{\lambda_2^2 - k_2^2})]^{-1}$$

$$\cdot \{\hat{g}_1^{(1)}(\sigma) + \hat{g}_3^{(2)}(\sigma) + \hat{r}_{1,c}(\sigma, \lambda_2)\}\Big)(\lambda_1) \tag{3.8}$$

96

for $Im\,\lambda_1 \gtrless \mp q_1$, $0 < (Im\,\lambda_2)^2 < q_2^2$, $q_1^2 + q_2^2 < q^2$. Taking a second symmetrization w.r.t. λ_1 now, what corresponds to a double $\mathcal{F}$–cosine transform, gives a formula for $\hat{\Phi}_{1,cc}(\lambda_1, \lambda_2)$ which is compared with a direct double $\mathcal{F}$–cosinus representation formula for the Helmholtz equation in Q_1:

$$\hat{\Phi}_{1,cc}(\lambda_1, \lambda_2) = \frac{\hat{g}_1^{(1)}(\lambda_1) + \hat{g}_{2,c}^{(1)}(\lambda_2)}{\lambda_1^2 + \lambda_2^2 - k_1^2} \,. \tag{3.9}$$

The comparison of both formulae for $\hat{\Phi}_{1,cc}$ and in a similar way for $\hat{\Phi}_{j,cc}(\lambda_1, \lambda_2)$ for $j = 2, 3, 4$ leads to a four-by-four system of integral equations of the form

$$\hat{\vec{g}}_c(\xi) + \int_0^\infty \underline{\underline{K}}(\xi, \eta) \cdot \hat{\vec{g}}_c(\eta) d\eta = \hat{\vec{v}}_c(\xi)\,, \quad \xi \in \mathbf{R} \tag{3.10}$$

for the four-vector unknown transformed functions $\hat{\vec{g}}_c(\xi)$ where one of the equations looks like

$$\hat{g}_{2,c}^{(1)}(\xi) + \frac{2}{\pi} \cdot \frac{\rho_2 \sqrt{\xi^2 - k_1^2} \cdot \sqrt{\xi^2 - k_2^2}}{\rho_1 \sqrt{\xi^2 - k_1^2} + \rho_2 \sqrt{\xi^2 - k_2^2}} \tag{3.11}$$

$$\cdot \left(\int_{\eta=0}^\infty \frac{\hat{g}_{1,c}^{(1)}(\eta)\,d\eta}{\eta^2 + \xi^2 - k_1^2} - \int_{\eta=0}^\infty \frac{\hat{g}_3^{(2)}(\eta)\,d\eta}{\eta^2 + \xi^2 - k_2^2} \right) = \hat{v}_{2,c}(\xi)\,.$$

The matrix $\underline{\underline{K}}(\xi, \eta)$ has zeros for $j = \ell \equiv 0\,(mod2)$ (*"chess-board structure"*) and the system (3.10) therefore may easily be reduced to a two-by-two system by inserting the second and foruth equation, respectively, into the first and third one of (3.10). The solvability for small $|k_j - k_\nu|$ and $|\rho_j - \rho_\nu|$ has been shown by N. LATZ and the author [5] (1984) and for the octant case in $\mathbf{R}^3$ [6] (1986) using Banach's fixed point principle.

3.2 The multimedia Sommerfeld half-plane or wedge problem

If a half plane $\mathcal{S}$ ($\stackrel{\wedge}{=} x \geq 0$) is inserted at the interface of the first and fourth quadrant, i.e. $\mathcal{S} = \partial Q_1 \cap \partial Q_4$ with different materials in Q_2 and Q_3 additionally or if the first quadrant Q_1 is replaced by one having prescribed boundary data on the faces $\mathcal{S}_1$ and $\mathcal{S}_2$: $B_1[\Phi_{tot}] = 0$ on $\mathcal{S}_1$ and $B_1[\Phi_{tot}] = 0$ on $\mathcal{S}_2$ one arrives along the procedure which leads to the system (3.9) to a three-by-three system of integral equations for the unknown $\mathcal{F}$–transformed Cauchy-data that are prescribed on $\mathcal{S}_1$ and $\mathcal{S}_2$. In case of different boundary conditions on $\mathcal{S}_1$ and

$\mathcal{S}_2$ it is more convenient to formulate an equivalent six-by-six system of integral equations for the unknown Cauchy-data on the semi-infinite axes $\mathcal{S}_1 \ldots \mathcal{S}_4$, where on $\mathcal{S}_2, \mathcal{S}_3, \mathcal{S}_4$ always two unkwnon data in the case of Sommerfeld's half-plane have to be determined while in the case of a right-angled wedge at Q_1 this happens only on $\mathcal{S}_3$ and $\mathcal{S}_4$, and one datum is left on $\mathcal{S}_{1\pm}$ and on $\mathcal{S}_1, \mathcal{S}_2$, respectively. For equal media in Q_2, Q_3, Q_4 the problem – *exterior wedge scattering* – has been investigated in Sobolev trace spaces by the author, F.-O. SPECK and F.S. TEIXEIRA [18, 11] (1991,1992): These problems have also been treated by the Kantorovich-Lebedev-transform as has been displayed in the author's article [7] (1987). For the special case of a "two-media boundary-transmission problem" with Dirichlet-, Neumann- or mixed boundary conditions on the line $y = 0$ and transmission conditions for two different media in Q_1 and Q_2 for their interface $\mathcal{S}_2$ the present author and F. PENZEL, F.-O. SPECK, and F.S. TEIXEIRA [11] (1993) gave an explicit solution by reduction to Wiener-Hopf-Hankel integral equations with bounded, piecewise continuous symbols on $\dot{\mathbf{R}}_\xi$. The author, N. LATZ and J. SCHEURER [12], calculated (1993) the generalized eigenfunctions and they gave the spectral representation in this case with a D–boundary condition on $y = 0$.

4 Concluding Remarks

The problems which were displayed are centering around scalar wave diffraction mainly in two space dimensions with a harmonic time dependence. Many more general problems concerning scattering of vectorial electro-magnetic and elastodynamical waves are still unsolved particularly those which cannot be reduced to pairs of scalar problems as in the case of cylindrical geometries. In three dimensions half-planes have to be replaced by polygonal plane (and more general thin) screens. The canonical problem of the diffraction by a quarter-plane leads to a two-dimensional Wiener-Hopf problem which – in general – is still unsolved though there are results by the author and F.-O. SPECK [8] (1988). Th octant's and polyhedrons diffraction problems are only scratched at the surface if treated as well-posed Sobolev space boundary transmission problems. K. ROTTBRAND recently got (1993) interesting results w.r.t. parallel multi-crack problems in an isotropic linear elastic material via six-by-six blocks of Fourier symbol matrices corresponding to the very general boundary transmission conditions on the half-plane, showing the same structure as K. ROTTBRAND obtained (1992) for the Helmholtz case. Similar results, for one crack, were obtained for the simplest "pure boundary conditions" of D- and N–type (tractions!) for the linear thermoelastic equations by C. ERBE

(1993) and for the linear viscoelastic type for wedges by J. MARK (1993).

The most challenging problems still to be attacked are the fully time-dependent initial boundary-transmission problem in order to get a better insight into the asymptotic behavior of the solutions for small ($t \to 0+$) and long times ($t \to +\infty$) (*"limiting amplitude principle"*). The spectral analysis for the problems, particularly for the vectorial case, is only at its very beginning and will play an important role for the study (e.g. for possible blow up or bifurcations' behavior) of corresponding non-linear wave equations and hyperbolic (parabolic) systems.

References

[1] Costabel, M., Stephan, E.: A Direct Boundary Integral Equation Method for Transmission Problems, J. Math. Anal. Appl. **106** (1985), 367–413

[2] Costabel, M., Stephan, E.: Strongly elliptic boundary integral equations for electromagnetic transmission problems. Proceed. Roy. Soc. Edinburgh **109A**, 271–296 (1988)

[3] Latz, N.: Untersuchungen über ein skalares Übergangswertproblem aus der Theorie der Beugung elektromagnetischer Wellen an dielektrischen Keilen. Diss. U Saarbürcken 1968, 117S.

[4] Latz, N.: Wiener-Hopf-Gleichungen zu speziellen Ausbreitungsproblemen elektromagnetischer Schwingungen. Habil.-schrift TU Berlin 1974, 65S.

[5] Latz, N., Meister, N.: On the Transmission Problem of the Helmholtz Equation for Quadrants. Math. Meth. Appl. Sci. **6** (1984), 129–157

[6] Meister, E.: Integral Equations for the Fourier Transformed Boundary Values for the Transmission Problems for Right-Angled Wedges and Octants, Math. Meth. Appl. Sci.: **8** (1986), 182–205

[7] Meister, E.: Einige gelöste und ungelöste kanonische Probleme der mathematischen Beugungstheorie: Expos. Math. **5** (1987), 193–237

[8] Meister, E., Speck, F.-O.: A contribution to the quarter-plane problem indiffraction theory. J. Math. Anal. Appl. **130** (1988), 223–236

[9] Meister, E., Speck, F.-O.: The Explicit Solution of Elastodynamical Diffraction Problems. Z. f. Anal. u. ihre Anw., Bd. **8** (4) (1989), 307–328

[10] Meister, E., Penzel, F., Speck, F.-O., Teixeira, F.S.: Two media scattering problems in a half-space. In: Partial Differential Equations with Real Analysis. (H. Begehr and A. Jeffrey, eds.) 122–146, Longman, London 1992

[11] Meister, E., Speck, F.-O., Teixeira, F.S.: Wiener-Hopf-Hankel operators for some wedge diffraction problems with mixed boundary conditions. J. Integr. Eq. Appl. **4** (2), (1992), 229–255

[12] Meister, E., Latz, N., Scheurer, J.: Specral Analysis of a transmission problem for the Helmholtz equation on the half-space (to appear in Rendic di Matem., Rome, (1993). no. 4

[13] Meister, E., Penzel, F., Speck, F.-O., Teixeira, F.S.: Two Canonical Wedge Problems for the Helmholtz Equation. Preprint 3/93, March 1993, Rep. Mat. Inst. Sup. Técn. Lisboa, Portugal, 28S.

[14] Mohr, R.: Eine spektraltheoretische Behandlung eines Übergangproblems. Diss. U Stuttgart 1976, 103S.

[15] v. Petersdorff, T.: Boundary integral equations for mixed Dirichlet, Neumann and transmission problems. Math. Meth. Appl. Sci.: **11** (1989), 185–213

[16] v. Petersdorff, T.: Randwertprobleme der Elastizitätstheorie für Polyeder – Singularitäten und Approximation mit Randelementmethoden. Diss. TH Darmstadt, 1989, 133S.

[17] Rottbrand, K., Meister, E., Speck, F.-O.: Wiener-Hopf equations for Waves Scattered by a System of Parallel Sommerfeld Half Planes. Math. Meth. Appl. Sci. **14** (1991), 525–552

[18] Teixeira, F.S.: Diffraction by a rectangular wedge: Wiener-Hopf-Hankel formulation. Integral Oper. Th. **14** (1991), 436–455

[19] Weber, C.: Hilbertraummethoden zur Untersuchung der Beugung elektromagnetischer Wellen an Dielektrika. Diss. U Stuttgart 1977, 116S.

Erhard Meister
Fachbereich Mathematik
Technische Hochschule
Schloßgartenstr. 7
D–64289 Darmstadt
e-mail: db0b@mathematik.th-darmstadt.de

Mathematical Problems of the Anisotropic Elasticity for Piece-wise Homogeneous Bodies

David Natroshvili

The investigation deals with the problems of anisotropic elasticity for composed bodies i.e. for bodies which have piece-wise homogeneous structure. The most general case of the structure of the elastic body under consideration mathematically can be described by the following way. In the three-dimensional Euclidian space $\mathbb{R}^3$ we have some closed smooth two-dimensional connected surface S_1 which involves other closed smooth surfaces $S_2, ... S_m (S_j \cap S_k = \vee, j \neq k)$. By these surfaces the space $\mathbb{R}^3$ is devided into several connected domains $\Omega_1, ..., \Omega_N$. Each domain Ω_l is supposed to be filled by an anisotropic material with corresponding different elastic coefficients

$$\overset{l}{c}_{kjpq} = \overset{l}{c}_{pqkj} = \overset{l}{c}_{jkpq}, \quad l = 1, 2, ..., N; \quad k, j, p, q = 1, 2, 3. \tag{1}$$

Common boundaries of two different materials are called contact boundaries of the piece-wise homogeneous body. If some domains represent empty inclusions then corresponding to them surfaces and S_1 are called boundary surfaces of the piece-wise homogeneous body.

Such piece-wise homogeneous structures encounter in many physical, mechanical and engineering applications.

Classical and non-classical mathematical problems for isotropic piece-wise homogeneous bodies by means of potential methods are studied in [KuGBB1], [Je1], [Ku1], while the similar problems for anisotropic piece-wise homogeneous bodies are investigated in [BuGe1], [JeNa1].

In the present paper we treat mixed boundary-contact problems. For simplicity we consider the following model problems (for domains of general structure described above all problems can be investigated similarly with slight modifications).

We suppose that the piece-wise homogeneous anisotropic body consists of two connected domains $\Omega_1 = \Omega^+$ and $\Omega_2 = \Omega^-$, provided Ω^+ to be bounded (diam $\Omega^+ < \infty$) and $\Omega^- = \mathbb{R}^3 \setminus \overline{\Omega^+}$. Thus we have only one contact surface $S = \partial\Omega^+ = \partial\Omega^-$.

Let a smooth, connected, non-intersecting curve $\gamma \in S$ devides the surface S into two parts S_1 and $S_2 : S = S_1 \cup S_2 \cup \gamma; \ \overline{S_l} = S_l \cup \gamma, l = 1, 2.$

The basic mixed contact (interface) problems can be formulated in the following way.

C-D Problem. Find vectors $\overset{1}{u}$ and $\overset{2}{u}$ satisfying in Ω_1 and Ω_2 respectively equations

$$\overset{l}{A}(D)\,\overset{l}{u}(x) = 0, \quad x \in \Omega_l, \tag{2}$$

and conditions on contact surface S

$$[\overset{1}{u}(x)]^+ - [\overset{2}{u}(x)]^- = f(x), \tag{3}$$

$$[\overset{1}{T}(D, n(x))\,\overset{1}{u}(x)]^+ - [\overset{2}{T}(D, n(x))\,\overset{2}{u}(x)]^- = F(x), \tag{4}$$

$$x \in S_1,$$

$$[\overset{1}{u}(x)]^+ = f_+(x), \tag{5}$$

$$[\overset{2}{u}(x)]^- = f_-(x), \tag{6}$$

$$x \in S_2,$$

where

$$\overset{l}{A}(D) = \|\,\overset{l}{c}_{kjpq}\,D_j D_q\|_{3\times 3}, \quad D_j = \frac{\partial}{\partial x_j}, \quad j = 1, 2, 3, \tag{7}$$

is a formally self-adjoint strongly elliptic matrix differential operator generated by statical equilibrium equations, $n(x) = (n_1(x), n_2(x), n_3(x))$ is the unit exterior normal vector at $x \in S, \overset{l}{u} = (\overset{l}{u}_1, \overset{l}{u}_2, \overset{l}{u}_3)^T$ is the displacement vector, $\overset{l}{T}(D, n(x)\,\overset{l}{u}(x)$ is the stress vector calculated on the surface element with the unit normal vector $n(x)$

$$[\overset{l}{T}(D, n(x))\,\overset{l}{u}(x)]_k = \overset{l}{c}_{kjpq}\,n_j(x)D_q\,\overset{l}{u}_p(x), \quad k = 1, 2, 3,$$

102

(here and in what follows summation over repeated indices is meant from 1 to 3), the symbol $[\cdot]^{\pm}$ designates limiting values on S from $\Omega^{\pm}$, $[\cdot]^T$ denotes transposition; F, f and $f_{\pm}$ are given vector functions.

In addition we suppose that

$$\overset{2}{u}(x) = o(1) \text{ as } |x| \to +\infty. \tag{8}$$

C-N Problem. Find vectors $\overset{1}{u}$ and $\overset{2}{u}$ satisfying equation (2) in Ω_1 and Ω_2, respectively, contact conditions (3), (4) on S_1 and

$$\left.\begin{aligned}
[\,\overset{1}{T}(D_x, n(x))\,\overset{1}{u}(x)]^+ &= F_+(x), \\
[\,\overset{2}{T}(D_x, n(x))\,\overset{2}{u}(x)]^- &= F_-(x),
\end{aligned}\right\} \quad \text{on } S_2.$$

$$\tag{9}$$
$$\tag{10}$$

The decay condition (8) is required as above.

We consider the both mixed problems in the Sobolev spaces

$$\overset{1}{u} \in W_p^1(\Omega^+), \quad \overset{2}{u} \in W_{p,loc}^1(\Omega^-). \tag{11}$$

The conditions (3)-(6) and (9)-(10) are understood in the sense discribed in [DNS1]. Therefore we provide the following conditions to be fulfilled

$$\begin{aligned}
f \in B_{p,p}^{1-1/p}(S_1), \quad &f_{\pm} \in B_{p,p}^{1-1/p}(S_2), \\
F \in B_{p,p}^{-1/p}(S_1), \quad &F_{\pm} \in B_{p,p}^{-1/p}(S_2),
\end{aligned} \tag{12}$$

$$f^0 = \left\{\begin{array}{ll} f & \text{on } S_1 \\ f_+ - f_- & \text{on } S_2 \end{array}\right\} \in B_{p,p}^{1-1/p}(S), \tag{13}$$

$$F^0 = \left\{\begin{array}{ll} F & \text{on } S_1 \\ F_+ - F_- & \text{on } S_2 \end{array}\right\} \in B_{p,p}^{-1/p}(S). \tag{14}$$

Here and in what follows by $B_{p,q}^{\nu}(\Omega^+)$, $B_{p,q,loc}^{\nu}(\Omega^-)$, $B_{p,q}^{\nu}(S)$ and $H_p^{\nu}(\Omega^+)$, $H_{p,loc}^{\nu}(\Omega^-)$, $H_p^{\nu}(S)$ are denoted the Besov and the Bessel potential spaces with $\nu \in \mathbb{R}$, $1 < p < \infty$, $1 \le q \le \infty$ (see [Tr1], [Tr2]).

We have also the following notations

$$
\begin{aligned}
B^\nu_{p,q}(S_l) &= \{f|_{S_l} : f \in B^\nu_{p,q}(S)\}, \\
\tilde{B}^\nu_{p,q}(S_l) &= \{f \in B^\nu_{p,q}(S) : \mathrm{supp} f \subset \overline{S}_l\}, \\
H^\nu_{p,q}(S_l) &= \{f|_{S_l} : f \in H^\nu_p(S)\}, \\
\tilde{H}^\nu_p(S_l) &= \{f \in H^\nu_p(S) : \mathrm{supp} f \subset \overline{S}_l\}
\end{aligned}
$$

(cf. [CS1], [DNS1]); by $f|_{S_l}$ is denoted the restriction on S_l.

Besides, for simplicity we suppose that S and γ possess C^∞ smoothness (in fact some finite regularity is sufficient). Let $\overset{l}{\Gamma} = \| \overset{l}{\Gamma}_{kj} \|_{3\times 3}$ be the fundamental matrix of the operator $\overset{l}{A}$ (see [Na1]) and introduce single- and double layer potentials

$$
(\overset{l}{V} g)(x) = \int_S \overset{l}{\Gamma}(x - y)g(y)dS_y, \tag{15}
$$

$$
(\overset{l}{U} g)(x) = \int_S [T(D_y, n(y)) \, \overset{l}{\Gamma}(y - x)]^T g(y)dS_y \tag{16}
$$

In the sequel we need the following operators generated by the above potentials

$$
(\overset{l}{\mathcal{H}} g)(x) = \int_S [\overset{l}{\Gamma}(x - y)g(y)dS_y, \quad x \in S, \tag{17}
$$

$$
(\overset{l}{\mathcal{K}} g)(x) = \int_S [T(D_x, n(x)) \, \overset{l}{\Gamma}(x - y)]g(y)dS_y, \quad x \in S, \tag{18}
$$

$$
(\overset{l}{\mathcal{K}}^* g)(x) = \int_S [T(D_y, n(y)) \, \overset{l}{\Gamma}(y - x)]^T g(y)dS_y, \quad x \in S, \tag{19}
$$

$$
(\overset{l}{\mathcal{L}} g)(x) = \lim_{\Omega^\pm \ni z \to x \in S} \overset{l}{T}(D_z, n(z))(\overset{l}{U} g)(z), \quad x \in S. \tag{20}
$$

104

Properties of these operators and potentials (15), (16) are studied in [Na1], [DNS1]. Applying the results obtained in above cited works we can prove

Theorem 1. If $\overset{1}{u} \in C^1(\overline{\Omega^+}) \cap C^2(\Omega^+)$ and $\overset{2}{u} \in C^1(\overline{\Omega^-}) \cap C^2(\Omega^-)$ satisfy equation (2) in $\Omega_1 = \Omega^+$ and $\Omega_2 = \Omega^-$, respectively, and

$$\left.\begin{array}{rcl} [\overset{1}{u}(x)]^+ - [\overset{2}{u}(x)]^- &=& \varphi(x), \\[2mm] [\overset{1}{T}(D,n(x))\,\overset{1}{u}(x)]^+ - [\overset{2}{T}(D,n(x))\,\overset{2}{u}(x)]^- &=& \Phi(x), \end{array}\right\} \quad x \in S, \quad (21)$$

with some given vector-functions φ and Φ, then the following representations

$$\overset{1}{u}(x) = \overset{1}{V}[\overset{1}{\mathcal{H}}{}^{-1} \mathcal{N}^{-1}(\Phi + \overset{2}{\mathcal{N}}\varphi)](x), \quad x \in \Omega^+, \tag{22}$$

$$\overset{2}{u}(x) = \overset{2}{V}[\overset{2}{\mathcal{H}}{}^{-1} \mathcal{N}^{-1}(\Phi - \overset{1}{\mathcal{N}}\varphi)](x), \quad x \in \Omega^-, \tag{23}$$

hold, where $\overset{l}{\mathcal{H}}{}^{-1}$ and $\mathcal{N}^{-1}$ are the inverse operators to $\overset{l}{\mathcal{H}}$ and $\mathcal{N}$, respectively,

$$\mathcal{N} = \overset{1}{\mathcal{N}} + \overset{2}{\mathcal{N}}, \tag{24}$$

$$\overset{1}{\mathcal{N}} = (-\frac{1}{2}\mathcal{I} + \overset{1}{\mathcal{K}})\,\overset{1}{\mathcal{H}}{}^{-1}, \quad \overset{2}{\mathcal{N}} = -(\frac{1}{2}\mathcal{I} + \overset{2}{\mathcal{K}})\,\overset{2}{\mathcal{H}}{}^{-1}, \tag{25}$$

$\mathcal{I}$ is the identity operator. If $\varphi = \Phi = 0$, then $\overset{1}{u} = 0$ and $\overset{2}{u} = 0$.

If $\varphi \in C^{\kappa+1+\alpha}(S)$ and $\Phi \in C^{\kappa+\alpha}(S)$, where $\kappa \geq 0$ is an integer and $0 < \alpha < 1$, then $\overset{l}{u} \in C^{\kappa+1+\alpha}(\overline{\Omega_l}), l = 1, 2$.

If $\varphi \in B_{p,p}^{\nu+1}(S)[B_{p,q}^{\nu+1}(S)], \quad \Phi \in B_{p,p}^{\nu}(S)[B_{p,q}^{\nu}(S)]$, then

$\overset{1}{u} \in H_p^{\nu+1+1/p}(\Omega^+)[B_{p,q}^{\nu+1+1/p}(\Omega^+)],$

$\overset{2}{u} \in H_{p,loc}^{\nu+1+1/p}(\Omega^-)[B_{p,q,loc}^{\nu+1+1/p}(\Omega^-)],$

$\nu \in \mathbf{R}, \ 1 < p < \infty, \ 1 \leq q \leq \infty.$ ∎

Theorem 2. The operators $\overset{1}{\mathcal{N}}, \overset{2}{\mathcal{N}}$ and $\mathcal{N}$ are singular integro-differential operators and have the following properties:

i) $\quad \overset{1}{\mathcal{N}}\mathcal{N}^{-1}\overset{2}{\mathcal{N}} = \overset{2}{\mathcal{N}}\mathcal{N}^{-1}\overset{1}{\mathcal{N}}$ and for any $f, g, h \in C^{1+\alpha}(S)$

$$(\overset{1}{\mathcal{N}} f, f)_{L^2(S)} \geq 0, (\overset{2}{\mathcal{N}} g, g)_{L^2(S)} \geq 0, (\mathcal{N}h, h)_{L^2(S)} \geq 0$$

with equality only for $g = h = 0$ and $f(x) = [a \times x] + b$, where a and b are arbitrary three-dimensional constant vectors, $[\cdot \times \cdot]$ stands for vector product and $(\cdot, \cdot)_{L^2(S)}$ stands for scalar product in $L^2(S)$;

ii) the operators $\overset{l}{\mathcal{N}}$ and $\mathcal{N}$ are formally self-adjoint pseudo-differential operators of order 1 with positive definite principal matrix symbols;

iii) the operators

$$\mathcal{N} : C^{\kappa+1+\alpha}(S) \to C^{k+\alpha}(S)$$

$$\mathcal{N} : H_p^{\nu+1}(S) \to H_p^{\nu}(S)[B_{p,q}^{\nu+1}(S) \to B_{p,q}^{\nu}(S)]$$

$$1 < p < \infty, 1 \leq q \leq \infty, \nu \in \mathbf{R},$$

are invertible. $\qquad\blacksquare$

Let us now go over to investigation of C-D Problem. First we change conditions (5) and (6) by equivalent equalities

$$\left. \begin{aligned} [\overset{1}{u}(x)]^+ - [\overset{2}{u}(x)]^- &= f_+(x) - f_-(x), \\ [\overset{1}{u}(x)]^+ + [\overset{2}{u}(x)]^- &= f_+(x) + f_-(x), \end{aligned} \right\} \quad x \in S_2. \qquad \begin{aligned} &(26) \\ &(27) \end{aligned}$$

Due to (13) it is evident that the difference

$$[\overset{1}{u}(x)]^+ - [\overset{2}{u}(x)]^- = f^0(x), \quad x \in S \tag{28}$$

is a known vector on S and belongs to the space $B_{p,p}^{1-1/p}(S)$. Let $\tilde{F}$ be some fixed extension of vector F from S_1 onto S_2 preserving the space

$$\tilde{F} \in B_{p,p}^{-1/p}(S), \quad \tilde{F}|_{S_1} = F.$$

Any other extension Φ of the vector F from S_1 onto S_2 preserving the space can be represented in the form

$$\Phi = \tilde{F} + \Psi \in B_{p,p}^{-1/p}(S)$$

with arbitrary

$$\Psi \in \tilde{B}_{p,p}^{-1,p}(S_2).$$

Let us now look for the solution of the C-D Problem in the form (cf.(22), (23))

$$\overset{1}{u}(x) = \overset{1}{V}[\overset{1}{\mathcal{H}}{}^{-1}\mathcal{N}^{-1}(\tilde{F} + \Psi + \overset{2}{\mathcal{N}}f^0)](x), \quad x \in \Omega_1, \tag{29}$$

$$\overset{2}{u}(x) = \overset{2}{V}[\overset{2}{\mathcal{H}}{}^{-1}\mathcal{N}^{-1}(\tilde{F} + \Psi - \overset{1}{\mathcal{N}}f^0)](x), \quad x \in \Omega_2, \tag{30}$$

with known vector-functions $f^0, \tilde{F}$ and sought for vector function Ψ.

It is evident that equations (2) and contact conditions (3), (4) and (26) are satisfied. Condition (27) leads for Ψ to the following pseudodifferential equation

$$r_2\mathcal{N}^{-1}\Psi = q \text{ on } S_2, \tag{31}$$

where r_2 is the restriction operator on S_2 and

$$q = \frac{1}{2}(f_+ + f_-) + r_2 Q \in B_{p,p}^{1-1/p}(S_2),$$

with

$$Q = -\mathcal{N}^{-1}\tilde{F} - \frac{1}{2}\mathcal{N}^{-1}(\overset{2}{\mathcal{N}} - \overset{1}{\mathcal{M}})f^0.$$

We have to investigate the solvability of (31) in the space $\tilde{B}_{p,p}^{-1/p}(S_2)$. To this end denote by $\sigma(x;\xi), x \in S, \xi \in \mathbf{R}^2$, the principal symbol matrix of operator $\mathcal{N}^{-1}$. Entries of $\sigma(x;\xi)$ are homogeneous functions of order -1 with respect to ξ.

To establish the Fredholm property (Noethericity) of equation (31) due to the general theory of pseudodifferential equations on manifold with boundary in the Besov and the Bessel potential spaces we have to investigate eigenvalues of the following matrix

$$M(x) = [\sigma(x; 0, -1)]^{-1}[\sigma(x; 0, 1)]$$

(see [Du1], [Sha1], [Sha2], [Esk1], [DNS1], [NCS1]).

Theorem 3. The eigenvalues of the matrix M for any $x \in S$ are positive numbers. ∎

By the method developed in [DNS1], [NCS1] we can prove

Theorem 4. Let $1 < p < \infty, 1 \leq q \leq \infty$. Then the operators

$$\mathcal{N}^{-1} : \tilde{B}^{\nu}_{p,q}(S_2) \to B^{\nu+1}_{p,q}(S_2), \tag{32}$$

$$\mathcal{N}^{-1} : \tilde{H}^{\nu}_p(S_2) \to H^{\nu+1}_p(S_2), \tag{33}$$

are bounded for all $\nu \in \mathbb{R}$.

Operator (32) is a Fredholm operator if the inequality

$$\frac{1}{p} - \frac{3}{2} < \nu < \frac{1}{p} - \frac{1}{2} \tag{34}$$

holds. Operator (33) is a Fredholm operator if and only if condition (34) holds. Both operators (32) and (33) are invertible if ν satisfies inequality (34). ∎

Theorem 4 implies uniquely solvability of equation (31) for

$$\frac{4}{3} < p < 4. \tag{35}$$

Consequently we obtain the following

Theorem 5. Let the conditions (12), (13) and (35) be fulfilled. Then the C-D Problem has unique solution satisfying (11). The solution is representable in the form (29), (30) with Ψ defined by uniquely solvable pseudodifferential equation (31). ∎

We can improve the regularity property of the solution by increasing smoothness of the boundary data. In particular the following theorem holds (cf. [DNS1], [NCS1]).

Theorem 6. Let $4/3 < p < 4$, $1 < t < \infty$, $1 \leq q \leq \infty$, $1/t - 3/2 < \nu < 1/t - 1/2$ and $\overset{1}{u}, \overset{2}{u}$ be the solution to the C-D Problem satisfying (11).

i) In addition, if

$$f \in B_{t,t}^{\nu+1}(S_1), f_{\pm} \in B_{t,t}^{\nu+1}(S_2), F \in B_{t,t}^{\nu}(S_1), f^0 \in B_{t,t}^{\nu+1}(S),$$

where f^0 is defined by (13), then

$$\overset{1}{u} \in H_t^{\nu+1+1/t}(\Omega_1), \overset{2}{u} \in H_{t,loc}^{\nu+1+1/t}(\Omega_2).$$

ii) If

$$f \in B_{t,q}^{\nu+1}(S_1), f_{\pm} \in B_{t,q}^{\nu+1}(S_2), F \in B_{t,q}^{\nu}(S_1), f^0 \in B_{t,q}^{\nu+1}(S),$$

then

$$\overset{1}{u} \in B_{t,q}^{\nu+1+1/t}(\Omega_1), \overset{2}{u} \in B_{t,q}^{\nu+1+1/t}(\Omega_2).$$

iii) In particular, if

$$f \in C^{\alpha}(S_1), f_{\pm} \in C^{\alpha}(S_2), F \in B_{\infty,\infty}^{\alpha-1}(S_1), f^0 \in C^{\alpha}(S),$$

then

$$\overset{1}{u} \in C^{\beta}(\overline{\Omega}_1), \overset{2}{u} \in C^{\beta}(\overline{\Omega}_2$$

for any $\beta \in (0, \alpha_0)$, $\alpha_0 = \min(\alpha, 1/2)$. $\blacksquare$

The problem C-N can be investigated by the same way. In this case the analoguous theorems are valid and the solution to the C-N Problem possesses the same regularity properties as the solution to the C-D problem provided that the boundary data satisfy corresponding conditions.

Remark 7. From the particular problems of mathematical physics and mechanics it is well known that, in general, solutions to mixed boundary value problems or their derivatives have singularities near the curves of discontinuity of boundary conditions (curves like γ) and they do not belong to the class of regular functions $C^1(\overline{\Omega^{\pm}})$. More over even for C^{∞} regular data of problems they do not posses even C^{α}-smoothness with $\alpha > 1/2$ in the vicinity of γ while being infinitely differentiable elsewere (i.e. in $\overline{\Omega^{\pm}} \setminus \gamma$).

Remark 8. The C-D and C-N Problems as particular cases involve screen and crack type problems for domains with interior cuts. In fact, if both elastic materials in Ω_1 and Ω_2 are the same and conditions (3) and (4) are homogeneous ($f = F = 0$), then the surface S_1, becomes a formal interface and disappears. Consequently the displacement vector satisfies equation (2) (which is the same for Ω_1 and Ω_2) for points $x \in S_1$ and we obtain the screen type problem (with conditions (5),(6)) or crack type problem (with conditions (9), (10)) for $\mathbf{R}^3$ with cut along the surface S_2.

References

[BuGe1] Burchuladze, T.V., Gegelia, T.G.: The Development of the Potential Methods in the Elasticity Theory. Metsniereba, Tbilisi, 1985.

[CS1] Costabel, M., Stephan, E.P.: An improved boundary element Galerkin method for three-dimensional crack problems. Integral Equations and Operator Theory. 10, 1987, 467-504.

[Du1] Duduchava, R.: On multidimensional singular integral operators. I, II. J. Operator Theory, 11, 1984, 41-76.

[DNS1] Duduchava, R., Natroshvili, D., Shargorodsky, E.: Boundary value problems of the mathematical theory of cracks. Proc. Inst. Appl. Math. Tbilisi Univ., 39, 1990, 68-84.

[Esk1] Eskin, G.I.: Boundary Value Problems for Elliptic Pseudodifferential Equations. Transl. of Math. Monogr., AMS, V. 52. Providence, Rhode Island, 1981.

[Je1] Jentsch, L.: Zur Existenz von regulren Lsungen der Elastostatik stckweise homogener Krper mit neuen Kontaktbedingungen an den Trennflchen zwischen zwei homogenen Teilen. Berlin: Akademie-Verlag, Abh. d. Schs. Akademie d. Wissenschaften zu Leipzig, Math. Naturw. Klasse. Bd. 53, H. 2, 1977.

[JeNa1] Jentsch, L., Natroshvili, D.: Non-classical interface problems for piecewise homogeneous anisotropic elastic bodies (to appear in Mathematical Methods in the Applied Sciences).

[Ku1] Kupradze, V.D.: Contact problems of the elasticity theory. Differential Equations, 16(2), 1980, 293-310.

[KuGBB1] Kupradze, V.D., Gegelia, T.G., Basheleishvili, M.O., Burchuladze, T.V.: Three-dimensional Problems of Mathematical Theory of Elasticity and Thermoelasticity. Nauka, Moscow, 1976.

[Na1] Natroshvili, D.: Investigation of boundary value and initial boundary value problems of mathematical theory of elasticity and thermoelasticity for homogeneous anisotropic bodies by means of potential methods. Doct. thesis. Math. Inst. Acad. Sci. GSSR, 1984, 1-325.

[NCS1] Natroshvili, D., Chkadua, O., Shargorodsky, E.: Mixed boundary value problems of the anisotropic Elasticity. Proc. Inst. Appl. Math. Tbilisi Univ., 39, 1990, 133-181.

[Sha1] Shargorodsky, E.: Boundary value problems for elliptic pseudodifferential operators: the half space case. Proc. Tbilisi Mathem. Inst. Acad. Sci. GSSR, 99, 1993, 44-80.

[Sha2] Shargorodsky, E.: Boundary value problems for elliptic pseudodifferential operators on manifolds. Proc. Tbilisi Mathem. Inst. Acad. Sci GSSR, 104, 1993.

[Tr1] Triebel, H.: Interpolation Theory, Function Spaces, Differential Operators. VEB Deutscher Verlag der Wissenschaften Berlin, 1978.

[Tr2] Triebel, H.: Theory of Function Spaces. Leipzig Birkhuser Verlag, Basel-Boston Stuttgart, 1983.

David Natroshvili
Georgian Technical University
Dept. of Mathematics(99)
M. Kostava str. 77

Tbilisi 380075
Georgia

Theory of Multipolar Fluids

Jindřich Nečas

1 Introduction

The most important impulse to the creation of the theory, is its ability to be a framework for the mathematical theory of compressible, heat conductive fluids. The theory provides proof of the existence of global (in time) solutions, what in spite of big efforts, the classical theory, based on linear Stoke's stress-strain relation, does not make possible. The theory is compatible with principles of thermodynamics and with the principle of material frame indifference. The physical theory of multipolar fluids appeared in the paper by Nečas, Šilhavý [1] and follows the general ideas of Green, Rivlin [2], [3]. The mathematical theory is developed in a serie of papers by Nečas, Novotný, Šilhavý [4], [5], [6], Nečas, Novotný [7], Nečas [8], Málek, Nečas, Růžička [9], [10], Bellout, Bloom, Nečas [11], [12], where also limits to monopolar, in general non-newtonian fluids are studied.

2 Physical theory

In details, see [1]. A multipolar fluid is given by the velocity field $v = (v_1, v_2, v_3)$, by the field of positive absolute temperature θ, by the density ρ, by specific internal energy e, by specific entropy η, by specific external body forces b, by the rate of external heat sources r, by the heat flux vector q and by spatial multipolar stress tensors $\tau_{ii_1,i_2...,i_kj}$, $k = 0, 1, \ldots, N - 1$, where the symmetry in indeces $i_1, \ldots, i_k$ is assumed. The above functions satisfy (summation on repeated indeces being understood)

$$(2.1) \quad \dot{\rho} + \rho \frac{\partial v_i}{\partial x_i} = 0 \qquad \text{(conservation of mass)},$$

$$(\cdot) = \frac{d}{dt} = \frac{\partial}{\partial t} + v_j \frac{\partial}{\partial x_j},$$

$$(2.2) \quad \rho \left(e + \tfrac{1}{2}|v|^2 \right) = \qquad \text{(conservation of energy)}$$

$$= \frac{\partial}{\partial x_i} \left(-q_i + \tau_{jj_1...j_ki} \frac{\partial^k v_j}{\partial x_{j_1} \cdots \partial x_{j_k}} \right) + \rho\, b_i\, v_i + \rho\, r,$$

112

$$(2.3) \quad \rho \, \dot{v}_i = \frac{\partial \tau_{ij}}{\partial x_j} + \rho \, b_i \qquad \text{(balance of momentum)},$$

$$(2.4) \quad \varepsilon_{ijk}\left(\tau_{jk} + \frac{\partial \tau_{ikp}}{\partial x_p}\right) = 0 \quad \text{(balance of angular momentum)},$$

$$(2.5) \quad \rho \, \dot{\eta} \geq -\frac{\partial}{\partial_t}\left(\frac{q_i}{\theta}\right) + \rho \frac{r}{\theta} \quad \text{(Clausius–Duhem inequality)}.$$

We suppose principal of material frame indifference: if a change of frame of the form:

$$(2.1) \qquad \overline{x} = Q_{ij}(t)\, x_j + c_i(t), \qquad Q_{ij}\, Q_{ik} = \delta_{jk},$$

then the principle of material frame indifference postulates that the scalars θ, ρ, e, r are invariant, while q_i, $\tau_{ij_1\ldots j_k j}$, $\frac{1}{2}\left(\frac{\partial v_i}{\partial x_j} + \frac{\partial v_j}{\partial x_i}\right) = e_{ij}$ and higher spatial derivatives of v change in the usual tensorial fashion.

We suppose constitutive relations

$$(2.2) \qquad e = e(\rho, \nabla v, \ldots, \nabla^K v, \theta, \nabla \theta)$$

$$(2.3) \qquad \tau_{ii_1\ldots i_k j} = \tau_{ii_1\ldots i_k j}(\rho, \nabla v, \ldots, \nabla^K v, \theta, \nabla \theta)$$

with $K = 2N - 1$.

It follows from the principle of material frame indifference, that in the constitutive relations the first spatial gradient of velocity appear only through its symmetric part.

Let us denote $\tau^E_{ii_1\ldots i_k j} = \tau_{ii_1\ldots i_k j}(\rho, 0, \ldots, 0, \theta, 0)$, $\tau^v = \tau - \tau^E$. We suppose $\tau^E_{ii_1\ldots i_p j} \equiv 0$ for $k \geq 1$, $q = -k \, \nabla \theta$, $k > 0$, $\tau_{ij} = \tau_{ji}$.

It follows from (2.5) with $\psi = e - \eta \, \theta$:

$$(2.4) \qquad e = e(\rho, \theta), \quad \eta = \eta(\rho, \theta), \quad \eta = -\frac{\partial \psi}{\partial \theta},$$

$$(2.5) \qquad \tau^E_{ij} = -p \, \delta_{ij}, \quad p = \rho^2 \frac{\partial \psi}{\partial \rho}, \quad \rho \, \dot{\psi} = -\rho \, \eta \, \dot{\theta} - p \frac{\partial v_i}{\partial x_i},$$

$$(2.6) \qquad \left(\tau^v_{jj_1\ldots j_k i} + \frac{\partial}{\partial x_p} \tau^v_{jj_1,\ldots,j_k i p}\right) \frac{\partial^{k+1} v_j}{\partial x_{j_1} \cdot \partial x_{j_k} \partial x_i} + k|\nabla \theta|^2 \frac{1}{\theta} \geq 0.$$

3   Tripolar compressible gas

For the mathematical theory of tripolar compressible, heat conductive gas, see [4], of several barotropic gas, see [6], or of isothermal gas, see [5]. There are in principle the same difficulties and mathematical methods are analogous. So we shall restrict ourselves to the last case where also a singular limit to dipolar gas is possible. We consider only weak solutions: in some cases, the singular limit to monopolar gas is possible, the solution being understood in the sense of Young measures, see [8]. Because the gas equation $p = R\rho\theta$, the constant positive temperature implies

$$(3.1) \qquad p = \alpha\rho, \quad \alpha > 0, \ \alpha \text{ positive constant.}$$

We suppose

$$(3.2) \qquad \tau_{ij}^{v} = \frac{\partial V}{\partial e_{ij}}(e) + L_{ij}(\nabla^3 v, , \nabla^5 v),$$

$$(3.3) \qquad \tau_{ii_1 j}^{v} = L_{ii_1 j}(\nabla^2 v, \nabla^4 v),$$

$$(3.4) \qquad \tau_{i_1,i_2 j}^{v} = L_{ii_1 i_2 j}(\nabla v, \nabla^3 v, \nabla^5 v)$$

where operators L are linear with constant coefficients (the dependence of odd respectively of even derivatives follow from the principle of material frame indifference).

We shall suppose also "Betti's formula", for v, $w \in C^\infty(\overline{\Omega}; \mathbb{R}^3)$, where Ω is a bounded domain with a boundary smooth enough, satisfying $v = w = 0$ on $\partial\Omega$:

$$(3.5) \qquad -\int_\Omega \frac{\partial \tau_{ij}^{v}}{\partial x_j} w_i \, dx = \int_\Omega \frac{\partial v}{\partial e_{ij}} e_{ij}(w) \, dx -$$

$$-\sum_{m=1}^{2} \int_{\partial\Omega} \tau_{ii_1\cdots i_m j}^{v} \frac{\partial^m w_i}{\partial x_{i_1}\ldots\partial x_{i_m}} \nu_j \, dS + ((v, w)).$$

Here $((v, w))$ is supposed symmetric, bounded, bilinear form, such that for $v = 0$ on $\partial\Omega$

$$(3.6) \qquad ((v, w)) \geq m\|v\|_{W^{3,2}}^2 \quad m > 0.[1]$$

3.7. Example. Dipolar fluid: $L_{ij} = -2\mu_1 \Delta e_{ij}$, $L_{ii_1 j} = 2\mu_1 \frac{\partial e_{ii_1}}{\partial x_j}$, tripolar fluid: $L_{ij} = -2\mu_1 \Delta e_{ij} + 2\mu_2 \Delta^2 e_{ij}$, $L_{ii_1 j} = 2\mu_1 \frac{\partial e_{ii_1}}{\partial x_j} - 2\mu_2 \Delta \frac{\partial e_{ii_1}}{\partial x_j}$, $L_{ii_1 i_2 j} = 2\mu_2 \frac{\partial^2 e_{ii_1}}{\partial x_{i_2} \partial x_j}$.

As far as for the viscous potential V, we suppose $V \in C^2(\mathbb{R}^9)$ and

$$(3.8) \qquad c_1(1 + |e|)^{p-2}|\xi|^2 \leq \frac{\partial^2 V}{\partial e_{ij} \partial e_{k\ell}} \xi_{ij}\, \xi_{k\ell} \leq c_2(1 + |e|)^{p-2}|\xi|^2.$$

So we have to solve in $Q_T := (0, T) \times \Omega$, $T < \infty$, the equations (2.1) and (2.3), with initial conditions $\rho(0, x) = \rho_0(x)$, $v(0, x) = v^0(x)$. Apart of non-slip conditions $v = 0$ on $\partial\Omega$, other 6 boundary conditions follow from the definition of the weak solution, which follows.

---

[1] $W^{k,p}$ are the usual Sobolev spaces. $W_0^{1,p}$ is the space with zero traces on $\partial\Omega$.

114

3.9. Definition. We look for $((0, T) = I)$

$$v \in L^\infty(I; W^{3,2}), \quad \frac{\partial v}{\partial t} \in L^2(I; L^2),$$

$$\rho \in L^\infty(I; W^{1,6}), \quad \frac{\partial \rho}{\partial t} \in L^\infty(I; L^6),$$

such that (2.1) is satisfied a. e. in Q_T, a. e. in I for any $w \in W^{3,2} \cap W_0^{1,2}$:

$$(3.10) \qquad \int_\Omega \rho \frac{\partial v_i}{\partial \tau} w_i \, dx + \int_\Omega \rho \, v_j \frac{\partial v_i}{\partial x_j} w_i \, dx$$

$$+ \; \alpha \int_\Omega \frac{\partial \rho}{\partial x_i} w_i \, dx + ((v, w)) + \int_\Omega \frac{\partial v}{\partial e_{ij}}(e(v)), e_{ij}(w) \, dx = 0$$

(zero body forces).

3.11. Theorem. Let $\rho_0 \in C^1(\overline{\Omega})$, $\rho_0 > 0$ in $\overline{\Omega}$, $v^0 \in W^{3,2} \cap W_0^{1,2}$. Then $\exists!$ to the problem 3.9.

I d e a o f t h e p r o o f. Let $\{w^k\}_{k=1}^\infty$ be an orthonormal basis in $W^{3,2}(\Omega; \mathbb{R}^3) \cap W_0^{1,2}(\Omega; \mathbb{R}^3)$ such that $((w^k, w^\ell)) = \delta_{k\ell}$. Let $c_i \in C^1(\overline{I})$ and put $v^m(t, x) = \sum_{i=1}^m c_i(t) \, w^i(x)$. Solve

$$(3.12) \qquad x^m(t) = v^m(t, x^m(t)), \quad x^m(0) = y \in \Omega.$$

Let $x = x^m(t, y)$ and put

$$(3.13) \qquad \rho_m(t, x) = \rho_0(y) \exp\left(- \int_0^t \frac{\partial v_i}{\partial x_i}(\tau, \, x^m(\tau)) \, d\tau \right).$$

ρ_m satisfies exactly (2.1).

Look for $\overline{v}^m(t, x) = \sum_{i=1}^m \overline{c}_i(t) \, w^i(x)$, such that $((\overline{v}^m(0), w^k)) = ((v^0, w^k))$, $k = 1, 2, \ldots, m$, and solve the "Osin's problem":

$$(3.14) \qquad \int_\Omega \left(\rho_m \frac{\partial \overline{v}^m}{\partial t} + \rho_m \, v_j^m \frac{\partial \overline{v}_i^m}{\partial x_j} + \alpha \frac{\partial \rho_m}{\partial x_i} \right) w_i^k \, dx$$

$$+ \; \int_\Omega \frac{\partial V}{\partial e_{ij}}(e(v^m)) \, e_{ij}(w^k) \, dx + ((\overline{v}^m, w^k)) = 0, \quad k = 1, 2, \ldots, m.$$

There exists a unique solution to the problem and for a short time, there exists a fixed point $\overline{v}^m = v^m$. But we have

$$(3.15) \qquad \frac{1}{2} \int_\Omega \rho_m(t) \, |v^m(t)|^2 \, dx - \frac{1}{2} \int_\Omega \rho_0 |v^m(0)|^2 \, dx$$

$$+ \int_0^t \int_\Omega \frac{\partial V}{\partial e_{ij}}(e(v^m))\, e_{ij}(v^m)\, d\tau\, dx +$$

$$+ \int_0^t ((v^m, v^m))\, d\tau + \alpha \int_\Omega \rho_m(t)\, (\ln \rho_m(t) - 1)\, dx -$$

$$-\alpha \int_\Omega \rho_0 (\ln \rho_0 - 1)\, dx = 0.$$

This implies the existence of v^m, ρ_m in I. $\qquad\square$

We have

3.16. Proposition. $\quad c_1 \exp\left(t^{-\frac{1}{2}}\gamma_1\right) \leq \rho_m(t, x) \leq c_2 \exp\left(t^{\frac{1}{2}}\gamma_1\right).$

3.17. Proposition.

$$\|\rho_m\|_{L^\infty(I; W^{1,2})} \leq c(\tau) < \infty, \qquad \left\|\frac{\partial \rho_m}{\partial t}\right\|_{L^\infty(I; L^2)} \leq c(\tau) < \infty.$$

From 3.16 and 3.17 we deduce more, testing (3.14) by $\frac{\partial v^m}{\partial t}$:

3.18. Proposition.

$$\|v_m\|_{L^\infty(I; W^{3,2})} \leq c(\tau),$$

$$\left\|\frac{\partial v^m}{\partial t}\right\|_{L^2(I; L^2)} \leq c(\tau),$$

$$\|\rho_m\|_{L^\infty(I; W^{1,6})} \leq c(\tau),$$

$$\left\|\frac{\partial \rho_m}{\partial t}\right\|_{L^\infty(I; L^6)} \leq c(\tau).$$

Taking finally v^{m_k}, an appropriate subsequence, and that $v^{m_k} \overset{*}{\rightharpoonup} v$ in $L^\infty(I; W^{3,2})$, $\frac{\partial v^{m_k}}{\partial t} \rightharpoonup \frac{\partial v}{\partial \tau}$ in $L^2(I; L^2)$, $\rho_{m_k} \overset{*}{\rightharpoonup} \rho$ in $L^\infty(I; W^{1,6})$ and $\frac{\partial \rho_{m_k}}{\partial \tau} \overset{*}{\rightharpoonup}$ in $L^\infty(I; L^6)$, we get the result.

4 Dipolar isothermal gas

In the case of the isothermal gas, using

$$(4.1) \qquad ((w, w)) \;=\; 2\mu_1 \int_\Omega \frac{\partial e_{ij}(v)}{\partial x_k} \frac{\partial e_{ij}(w)}{\partial x_k}\, dx +$$

$$+ 2\mu_2 \int_\Omega \frac{\partial^2 e_{ij}(v)}{\partial x_k \partial x_\ell} \frac{\partial^2 e_{ij}(w)}{\partial x_k \partial x_\ell}\, dx,$$

116

we can let $\mu_2 \to 0$. We suppose in (3.8): $1 \le p \le 2$.

We shall write the system (2.3) in the form

$$(4.2) \qquad \frac{\partial}{\partial t}(\rho v_i) + \frac{\partial}{\partial x_j}(\rho v_i v_j + \alpha \rho \delta_{ij} - \tau_{ij}^v) = 0.$$

Let $\psi(t) = (1+t)\ln(1+t) - t$, $\Phi(t) = e^t - t - 1$ and define the Orlicz space by

$$(4.3) \qquad \|u\|_{L_f} := \inf_{h>0}\left\{ h > 0; \int_\Omega f\left(\frac{|u(x)|}{u}\right)\, dx \le 1\right\}.$$

If $C_\Phi(\Omega)$ is the closure of $C(\overline{\Omega})$ in (4.3), we have:

$$(4.4) \qquad (C_\Phi(\Omega))^* = L_\psi(\Omega).$$

4.5 Theorem. Let the conditions of the theorem 3.11 be satisfied. Then there exists a solution to (2.1), (2.3) (in the sense of distributions), for $\mu_2 = 0$, such that

$$v \in L^2(I; W^{2,2} \cap W_0^{1,2}),$$
$$\rho \in L^\infty(I; L_\psi),$$
$$\frac{\partial \rho}{\partial t} \in L^2(I; W^{-3,2}), \quad ((W_0^{3,2})^* = W^{-3,2}),$$
$$\frac{\partial}{\partial t}(\rho v) \in L^2(I; W^{-3,2}).$$

Idea of the proof. We use the inequality (3.15) for v^k, ρ_k, solutions for $\mu_2^k > 0$. Let $\mu_2^k \to 0$. We can suppose $v^k \rightharpoonup v$ in $L^2(I; W^{2,2})$, $\rho_k \overset{*}{\rightharpoonup} \rho$ in $L^\infty(I; L_\psi)$, $\frac{\partial \rho_k}{\partial t} \rightharpoonup \frac{\partial \rho}{\partial t}$ in $L^2(I; W^{-3,2})$, $\frac{\partial}{\partial t}(\rho_k v^k) \rightharpoonup \frac{\partial}{\partial t}(\rho v)$ in $L^2(I; W^{-3,2})$. $\qquad \square$

5 Dipolar, incompressible, non-newtonian fluids and there singular limits to monopolar fluids

The detailed version of the problem can be found in papers [9], [10] and [12]. We shall restrict ourselves to the case 3.7. So with $\rho \equiv 1$, we have to solve the Navier–Stokes equation

$$(5.1) \qquad \frac{\partial v_i}{\partial t} + v_j \frac{\partial v_i}{\partial x_j} - \frac{\partial \tau_{ij}^v}{\partial x_j} + \frac{\partial p}{\partial x_i} = 0 \quad (b_i = 0).$$

We suppose for the exponent p from (3.8) $1 < p < 6$. Of course, the condition of incompressibility

$$(5.2) \qquad \frac{\partial v_i}{\partial x_i} = 0 \quad \text{in } Q_T$$

has to be satisfied.

By a standard method we get (use also the section 3)

5.3. Theorem. Let $v^0 \in W^{2,2}(\Omega; \mathbb{R}^3) \cap W_0^{1,2}(\Omega; \mathbb{R}^3)$, $\frac{\partial v_i^0}{\partial x_i} = 0$. $\exists!$ weak solution $w \in L^\infty(I; W^{2,2})$, $\frac{\partial v}{\partial t} \in L^2(I; L^2)$, such that

$$(5.4) \qquad \int_0^T |v|_{W^{1,p}}^p \, dt \le c,$$

$$(5.5) \qquad \operatorname*{ess\,sup}_I \|v(t)\|_{L^2(\Omega)} \le c,$$

where c is independent of μ_1.

Let us look for the space-periodic solution, provided $\int_{(0,L)^3} v^0 \, dx = 0$ and we look for v satisfying $\int_{(0,L)^3} v(t) \, dx = 0$.

If $\frac{6}{5} < p \le \frac{9}{5}$, we get as proved in [11] and [12] a measure valued solution.

Let us look for some estimates, which allow us to get weak solutions for monopolar fluids for $p > \frac{9}{5}$ and unique, regular week solutions for $p \ge \frac{11}{5}$.

Multiply (5.1) by $w_i = \frac{v_i''}{(1+\|v\|_{W^{1,2}}^2)^\lambda}$, where $v' = \frac{\partial v}{\partial x_\ell}$. For $\lambda \ne 1$,[1] we get

$$\frac{1}{2(1-\lambda)}\left(1+\|v(t)\|_{W^{1,2}}^2\right)^{1-\lambda} - \frac{1}{2(1-\lambda)}\left(1+\|v(0)\|_{W^{1,2}}^2\right)^{1-\lambda} +$$

$$+\ \beta \int_0^t \left(1+\|v(t)\|_{W^{1,2}}^2\right)^{-\lambda} dt \int_\Omega (1+|e|)^{p-2} e_{ij}(v')\, e_{ij}(v')\, dx\ +$$

$$+\ 2\mu_1 \int_0^t \left(1+\|v(t)\|_{W^{1,2}}^2\right)^{-\lambda} dt \int_\Omega \frac{\partial e_{ij}}{\partial x_k}(v') \frac{\partial e_{ij}}{\partial e_\ell}(v')\, dx\ \le$$

$$\le\ c_1 \int_0^t \left(1+\|v(t)\|_{W^{1,2}}^2\right)^{-\lambda} \|v(t)\|_{W^{1,3}}^3 \, dt.$$

For $p \ge \frac{11}{5}$, we get $\lambda \le 1$, hence independently of μ_1

$$(5.6) \qquad \sup_I \|v(t)\|_{W^{1,2}} \le c,$$

$$(5.7) \qquad \int_{Q_T} (1+|e|)^{p-2} e_{ij}\left(\frac{\partial v}{\partial x_\ell}\right) e_{ij}\left(\frac{\partial v}{\partial x_\ell}\right) dt dx \le c.$$

---

[1] For $\lambda = 1$ the first two terms are replaced by $\frac{1}{2}\ln(1+\|v(t)\|_{1,2}^2) - \frac{1}{2}\ln(1+\|v(0)\|_{1,2}^2)$.

For $2 \leq p < \frac{11}{5}$, with $\lambda = \frac{2(3-p)}{3p-5}$, we have

$$(5.8) \qquad \int_0^T \|v(t)\|_{W^{2,2}}^{2\theta}\, dt \leq c \int_0^T \left(1 + \|v(t)\|_{W^{1,2}}^2\right)^{-\lambda} \cdot$$

$$\int_\Omega (1 + |e|)^{p-2} \frac{\partial e_{ij}}{\partial x_k} \frac{\partial e_{ij}}{\partial x_k}\, dx \leq c,$$

with $\theta = \frac{p(3p-5)}{3(p^2-3p+4)}$ (independently of μ_1).

If $\frac{9}{5} < p < 2$, then with the same λ as above and with $\eta = \frac{p(10p-18)}{16p-2p^2-18}$ we get

$$(5.9) \qquad \int_0^T \|v(t)\|_{W^{2,p}}^{\eta}\, dt \leq c,$$

independent of μ_1. So we get

5.11. Theorem. Let the conditions to the theorem 5.3 be satisfied. Then for the space-periodic solution and $\mu_1 = 0$, there exists a unique weak solution to the problem (5.1), (5.2), such that $v \in L^p(I; W^{1,p})$ and (5.7), (5.8) are satisfied, provided $\frac{11}{5} \leq p$. If $\frac{9}{5} < p < \frac{11}{5}$, then there exists a weak solution, such that $v \in L^p(I; W^{1,p}) \cap L^\infty(I; L^2)$.

References

[1] Nečas, J., Šilhavý, M.: Multipolar viscous fluids. Quarterly of Applied Mathematics **XLIX** (1991), 2, 247–266.

[2] Green, A. E., Rivlin, R. S.: Simple force and stress multipoles. Arch. Rat. Mech. Anal. **16** (1964), 325–353.

[3] Green, A. E., Rivlin, R. S.: Multipolar continuum mechanics. Arch. Rat. Mech. Anal. **17** (1964), 113–147.

[4] Nečas, J., Novotný, A., Šilhavý, M.: Global solution to the ideal compressible multipolar heat conductive fluid. Comment. Math. Univ. Carol. **30** (1989), 3, 551–564.

[5] Nečas, J., Novotný, A., Šilhavý, M.: Global solution to the compressible isothermal multipolar fluid. J. Math. Anal. and Appl. **161** (1991), 1, 223–241.

[6] Nečas, J., Novotný, A., Šilhavý, M.: Global solution to the viscous compressible barotropic multipolar gas. Theoret. Comput. Fluid Dynamics **4** (1992), 1–11.

[7] Nečas, J., Novotný, A.: Some quantitative properties of the viscous compressible heat conductive multipolar fluid. Commun in P. E. D. **16** (1991), (2 & 3), 197–220.

[8] Nečas, J.: Theory of multipolar viscous fluids, the mathematics of finite elements and applications VII. In: MAFELAP 1990, Academic Press, 1991, 233–244.

[9] Málek, J., Nečas, J., Růžička M.: Measure-valued solutions and asymptotic behavior of a multipolar model of a boundary layers. Czech Math. J. **42** (1992), 117, 549–576.

[10] Málek, J., Nečas, J., Růžička, M.: On the non-newtonian incompressible fluids. Math. Models and Meth. in Appl. Sc., **3** (1993), 1, 35–63.

[11] Bellout, H., Bloom, F., Nečas, J.: Uniqueness and stability to the initial boundary value problem for bipolar viscous fluids. SIAM J. Math. Anal. **24** (1993), 1, 26–45.

[12] Bellout, H., Bloom, F., Nečas, J.: Phenomenological behavior of multipolar viscous fluids (to appear).

This work was partially supported by the grant 201/93/2177 of the Czech Grant Agency.

Jindřich Nečas
Charles University, Prague
Northern Illinois University, Dekalb

Domain Decomposition Methods for Singular Elliptic Problems

Sergeĭ V. Nepomnyashchikh

Abstract

This paper suggests a technique for the construction of preconditioning operators for the iterative solution of systems of grid equations approximating elliptic boundary value problems with strong singularities in the coefficients. The technique suggested is based on the decomposition of the original domain into subdomains in which the singularity of coefficients is characterized by some parameter. The convergence rate of the iterative process which uses the preconditioner suggested is independent of both the mesh size and the equation coefficients.

1 Introduction

In this paper, we design preconditioning operators for the system of grid equations approximating the following boundary value problem:

$$-\sum_{i,j=1}^{2} \frac{\partial}{\partial x_i}\left(a_{ij}(x)\frac{\partial u}{\partial x_j}\right) + a_0(x)u = f(x), \quad x \in \Omega ; \quad u(x) = 0 , \quad x \in \Gamma \quad (1)$$

We assume that Ω is a bounded, polygonal region and Γ is its boundary. Let Ω be a union of n nonoverlapping subdomains Ω_i,

$$\overline{\Omega} = \bigcup_{i=1}^{n} \overline{\Omega}_i, \quad \Omega_i \bigcap \Omega_j = \emptyset, \quad i \neq j,$$

where Ω_i are polygons and Γ_i are their boundaries. Let $\Omega^h = \bigcup_{i=1}^{n} \Omega_i^h$ be a regular triangulation of Ω which is characterized by a parameter h.
Let us introduce the weighted Sobolev spaces $H_\alpha^1(\Omega)$ with the norms as follows; see also [11].

$$\left.\begin{array}{rl}
\|u\|_{H_\alpha^1(\Omega)}^2 &= \|u\|_{L_2(\Omega)}^2 + |u|_{H_\alpha^1(\Omega)}^2 , \\[2mm]
\|u\|_{L_2(\Omega)}^2 &= \int_\Omega u^2(x)dx, \\[2mm]
|u|_{H_\alpha^1(\Omega)}^2 &= \int_\Omega \left(\frac{|\nabla u(x)|}{(\varrho(x))^{\alpha(x)}}\right)^2 dx .
\end{array}\right\} \quad (2)$$

Here we have $\alpha(x) \equiv \alpha_i = \text{const}$, $x \in \Omega_i$; and $\varrho(x)$ is the distance between the point $x \in \Omega_i$ and the boundary Γ_i of the subdomain Ω_i. We assume that $|\alpha| < \frac{1}{2}$. Denote by $\mathring{H}^1_\alpha(\Omega)$ the subspace of $H^1_\alpha(\Omega)$ with zero trace on Γ and introduce the bilinear form

$$a(u, v) = \int_\Omega \left(\sum_{i,j=1}^{2} a_{ij}(x) \frac{\partial u}{\partial x_j} \frac{\partial v}{\partial x_i} + a_0(x)uv \right) dx$$

and the linear functional $\quad l(v) = \int_\Omega f(x)vdx$.

We assume that the coefficients of the problem (1) are such that $a(u, v)$ is a symmetric, coercive and continuous form in $\mathring{H}^1_\alpha(\Omega) \times \mathring{H}^1_\alpha(\Omega)$. Denote by W a space of real–valued continuous functions linear on triangles of the triangulation Ω^h. A weak formulation of (1) is the following :

Find $u \in \mathring{H}^1_\alpha(\Omega)$ such that
$$a(u, v) = l(v), \quad \forall v \in \mathring{H}^1_\alpha(\Omega). \tag{3}$$

Using the finite element method we can pass from (3) to the linear algebraic system
$$Au = f. \tag{4}$$

The condition number of the matrix A depends on h, α and can be large enough. Our purpose is the design of a preconditioner B for the problem (4) such that the following inequalities are valid:

$$c_1(Bu, u) \le (Au, u) \le c_2(Bu, u), \quad \forall u \in R^N. \tag{5}$$

Here N is the dimension of W, the positive constants c_1, c_2 are independent of h and α, and the action of B^{-1} on a vector can be realized at low cost. To construct the preconditioning operator B, we use the additive Schwarz method [1,3,4] and the fictitious space method [5,10].

2 Additive Schwarz method for singular elliptic problems

The construction of the preconditioner for the system (4) will be realized on the basis of the additive Schwarz method [1,3,4]. To design the preconditioning operator B, we completely follow [7,10] and decompose the space W into the following sum of subspaces : $W = W_0 + W_1$.

To this end, divide the nodes of the triangulation Ω^h into two groups: those which lie inside of Ω_i^h and those which lie on boundaries of Ω_i^h. The subspace W_0 corresponds to the first set. We define $W_0 = \left\{u^h \in W \mid \ u^h(x) = 0, x \in S\right\}$, where

$$W_{0,i} = \{u^h \in W_0 \mid \ u^h(x) = 0, \quad x \overline{\in} \Omega_i^h\}, \quad i = 1, 2, \ldots, n.$$

It is clear that W_0 is the direct sum of the orthogonal subspaces $W_{0,i}$ with respect to the scalar product in $\mathring{H}_\alpha^1(\Omega)$, i.e., we have :

$$W_0 = W_{0,1} \oplus \ldots \oplus W_{0,n}.$$

The subspace W_1 corresponds to the second group of nodes Ω^h and can be defined in the following way. First, define V which is the space of traces of functions from W on S:

$$V = \{\varphi^h \mid \ \varphi^h(x) = u^h\big|_S, \quad u^h \in W\}.$$

To define the subspace W_1, we need a norm preserving extension operator of functions given at S into Ω^h. The basis of the further construction is the following trace theorem for the weight Sobolev spaces $H_\alpha^1(\Omega)$; see also [11]. We introduce the following denotations :

$$\left.\begin{aligned}
\|\varphi\|^2_{H^{\frac{1}{2}+\alpha}(\Gamma)} &= \|\varphi\|^2_{L^2(\Gamma)} + |\varphi|^2_{H^{\frac{1}{2}+\alpha}(\Gamma)}, \\
\|\varphi\|^2_{L^2(\Gamma)} &= \int_\Gamma \varphi^2(x)dx, \\
|\varphi|^2_{H^{\frac{1}{2}+\alpha}(\Gamma)} &= \int_\Gamma \int_\Gamma \frac{(\varphi(x)-\varphi(y))^2}{|x-y|^{2+2\alpha}} \, dx\,dy \,, \quad \text{and} \\
H(S) &= \left\{\varphi \mid \ \varphi\big|_{\Gamma_i} = \varphi_i, \quad \varphi_i \in H^{\frac{1}{2}+\alpha_i}(\Gamma_i)\right\}, \\
\|\varphi\|^2_{H(S)} &= \sum_{i=1}^n \|\varphi\|^2_{H^{\frac{1}{2}+\alpha_i}(\Gamma_i)} \,.
\end{aligned}\right\} \quad (6)$$

To define the bounded extension operator for the finite element case from V into W, we need mesh counterparts of the norms (2) and (6). To this end, let us split the triangles $\mathcal{T}_j$ of the triangulation Ω^h into three groups. Denote by M_1 a set of such $\mathcal{T}_j$ that $\mathcal{T}_j$ do not have vertices on S, denote by M_2 a set of such $\mathcal{T}_j$ that $\mathcal{T}_j$ have only one vertex on S, and denote by M_3 a set of such $\mathcal{T}_j$ that $\mathcal{T}_j$ have more than one vertex on S. Set

$$\begin{aligned}
\|u^h\|^2_{H_{\alpha,h}^1(\Omega)} &= \sum_{i=1}^n \|u^h\|^2_{H_{\alpha_i,h}^1(\Omega_i)}, \\
\|u^h\|^2_{H_{\alpha_i,h}^1(\Omega_i)} &= \|u^h\|^2_{L_{2,h}(\Omega_i)} + |u^h|^2_{H_{\alpha_i,h}^1(\Omega_i)}, \\
\|u^h\|^2_{L_{2,h}(\Omega_i)} &= \sum_{z_j \in \Omega_i^h} (u^h(z_j))^2 h^2 \,, \quad \text{where}
\end{aligned}$$

$$|u^h|^2_{H^1_{\alpha_i,h}(\Omega_i)} = \sum_{T_j \in M_1 \cap \Omega_i} \frac{(u_{j1}-u_{j2})^2+(u_{j2}-u_{j3})^2+(u_{j3}-u_{j1})^2}{(\varrho(T_j,\gamma))^{\epsilon\alpha_i}} +$$
$$+ \sum_{T_j \in M_2 \cap \Omega_i} \frac{(u_{j1}-u_{j2})^2+(u_{j2}-u_{j3})^2+(u_{j3}-u_{j1})^2}{h^{2\alpha_i}} +$$
$$+ \sum_{T_j \in M_3 \cap \Omega_i} \frac{(u_{j1}-u_{j2})^2+(u_{j2}-u_{j3})^2+(u_{j3}-u_{j1})^2}{(1-2\alpha_i)h^{2\alpha_i}}, \quad \forall u^h \in W.$$

Here z_i are vertices of Ω^h, u_{j1}, u_{j2}, u_{j3} are values of u^h at vertices of T_j, and $\varrho(T_j, \Gamma_i)$ is the distance between T_j and Γ_i.

Using the natural order of nodes on Γ_i, let us put for each node $z_j \in \Gamma_i$ into correspondence the node z_{j+1}, which is a node neighbouring upon z_i, and set

$$\|\varphi^h\|^2_{H_h(S)} = \sum_{i=1}^{n} \|\varphi^h\|^2_{H_h^{\frac{1}{2}+\alpha_i}(\Gamma_i)},$$
$$\|\varphi^h\|^2_{H_h^{\frac{1}{2}+\alpha_i}(\Gamma_i)} = \|\varphi^h\|^2_{L_{2,h}(\Gamma_i)} + |\varphi^h|^2_{H_h^{\frac{1}{2}+\alpha_i}(\Gamma_i)},$$
$$\|\varphi^h\|^2_{L_{2,h}(\Gamma_i)} = \sum_{z_j \in \Gamma_i} (\varphi^h(z_j))^2 h, \quad \text{where}$$
$$|\varphi^h|_{H_h^{\frac{1}{2}+\alpha_i}(\Gamma_i)} = \sum_{z_j \in \Gamma_i, j \neq k} \sum_{z_k \in \Gamma_i} \frac{(\varphi^h(z_j) - \varphi^h(z_k))^2}{|z_j - z_k|^{2+2\alpha_i}} h^2 +$$
$$+ \sum_{z_j \in \Gamma_i^h} \frac{(\varphi^h(z_j) - \varphi^h(z_{j+1}))^2}{(1-2\alpha_i)h^{2\alpha_i}}, \quad \forall \varphi^h \in V.$$

The following lemmata are valid.

Lemma 1 *There exist positive constants c_5 and c_6, independent of α and h, such that*
$$c_5\|u^h\|_{H^1_{\alpha_i}(\Omega_i)} \leq \|u^h\|_{H^1_{\alpha_i,h}(\Omega_i)} \leq c_6\|u^h\|_{H^1_{\alpha_i}(\Omega_i)}, \quad \forall u^h \in W, \quad i = 1, 2, \ldots, n.$$

Lemma 2 *There exist positive constants c_7 and c_8, independent of α and h, such that*
$$c_7\|\varphi^h\|_{H^{\frac{1}{2}+\alpha_i}(\Gamma_i)} \leq \|\varphi^h\|_{H_h^{\frac{1}{2}+\alpha_i}(\Gamma_i)} \leq c_8\|\varphi^h\|_{H^{\frac{1}{2}+\alpha_i}(\Gamma_i)}, \quad \forall \varphi^h \in V, \quad i = 1, 2, \ldots, n.$$

Let us use the explicit extension operator
$$t^h : V \to W , \tag{7}$$

which was suggested for regular elliptic second order problems. The definition and the realization algorithm were done in [5, 6, 8] and briefly can be described in the following way. Let us introduce the near–boundary coordinate system (s, n) which is defined in a δ–neighborhood of Γ_i. Here s defines a point P at Γ_i and n is the distance between the given point and Γ_i along the internal pseudo-normal, whose direction at the angular points coincides with the bisectrix of the angle and along the smooth part the vector changes, e.g., linearly.

Let us define

$$t : H(S) \to H_\alpha^1(\Omega) , \quad t\varphi = u ,$$

$$u(s,n) = \left(1 - \tfrac{n}{\delta}\right) \int\limits_{s}^{s+n} \tfrac{\varphi(t)}{n} \, dt , \qquad (8)$$

where the function u is extended by zero in the rest of Ω. Using the auxiliary mesh, which is topologically equivalent to a uniform rectangular mesh, we can define the finite element analogue t^h (7) of the operator t from (8). The following theorem is valid.

Theorem 1 *There exists a positive constant c_9, independent of α and h, such that*

$$\|u^h\|_{H_{\alpha,h}^1(\Omega)} = \|t^h\varphi^h\|_{H_{\alpha,h}^1(\Omega)} \le c_9 \|\varphi^h\|_{H_h^{\frac{1}{2}+\alpha}(S)}, \quad \forall \varphi^h \in V.$$

Remark :
The cost of the actions of t^h and $(t^h)^*$ on vectors is equivalent to $O(h^{-2})$ arithmetic operations (see [5] for details).

At last, we can define the subspace W_1 as $W_1 := \left\{ u^h \mid \; u^h = t^h\varphi^h, \; \varphi^h \in V \right\}$. It is obvious that $W = W_0 + W_1$ is valid and this decomposition of the space W is regular in the following sense.

Theorem 2 *There exists a positive constant c_{10}, independent of α and h, such that for any function $u^h \in W$ there exist $u_i \in W_i, i = 0, 1$, such that $u_0 + u_1 = u$ and $\|u_0\|_{H_\alpha^1(\Omega)} + \|u_1\|_{H_\alpha^1(\Omega)} \le c_{10}\|u\|_{H_\alpha^1(\Omega)}$ are fulfilled.*

3 Construction of equivalent norms on interfaces

According to [8] we can construct a preconditioner for the subspace W_1 in the following form : $B_1^+ = t^h \Sigma^{-1}(t^h)^*$, where Σ has to satisfy

$$c_1\|\varphi^h\|_{H(S)}^2 \le (\Sigma\varphi, \varphi) \le c_2\|\varphi^h\|_{H(S)}^2, \quad \forall \varphi^h \in V. \qquad (9)$$

Here the components of the vector φ are equal to the values of the function φ^h in corresponding nodes. The constants c_1, c_2 should be independent of α and h. The construction of the easily invertible operator (matrix) Σ is a goal of this section.

Let us start the construction of the preconditioner with the so-called strip case.

In this case the interface S is an union of nonoverlapping interfaces S_i

$$S = \bigcup_{i=1}^{m} S_i, \quad S_i \bigcap S_j = \emptyset, \quad i \neq j,$$

and each S_i is topologically equivalent to a segment. Let S_i be the interface between the subdomains Ω_{i_1} and Ω_{i_2}. It is easy to see that there exist positive constants c_3 and c_4, independent of α and h, such that

$$c_3\|\varphi^h\|_{H(S)} \leq \sum_{i=1}^{m} \|\varphi^h\|_{\hat{H}^{\frac{1}{2}+\beta_i}(S_i)} \leq c_4\|\varphi^h\|_{H(S)}, \quad \forall \varphi^h \in V .$$

Here we have $\beta_i = \max\{\alpha_{i_1}, \alpha_{i_2}\}$ and

$$\|\varphi^h\|^2_{\hat{H}^{\frac{1}{2}+\beta_i}(S_i)} = \begin{cases} \|\varphi^h\|^2_{H^{\frac{1}{2}+\beta_i}(S_i)}, & \beta_i \neq 0, \\ \|\varphi^h\|^2_{H^{\frac{1}{2}+\beta_i}(S_i)} + \int\limits_{S_i}(\varphi^h(x))^2 \left(\frac{1}{\varrho(x,a_i)} + \frac{1}{\varrho(x,b_i)}\right) \, dx, & \beta_i = 0, \end{cases}$$

where a_i and b_i are the extreme points of S_i. Set

$$\left. \begin{aligned} \Sigma_i &= h\left(\tfrac{1}{h^2}A_i\right)^{\frac{1}{2}+\beta_i}, \\ A_i &= \begin{pmatrix} 2 & -1 & & \\ -1 & \ddots & \ddots & \\ & \ddots & \ddots & -1 \\ & & -1 & 2 \end{pmatrix}, \quad i = 1,2,\ldots,m, \\ \Sigma &= \operatorname{diag}(\Sigma_1,\ldots,\Sigma_m). \end{aligned} \right\} \tag{10}$$

Here the dimension n_i of the matrix A_i is equal to the dimension of the finite element space $\hat{H}^{\frac{1}{2}+\beta_i}(S_i)$. By interpolation, the inequalities (9) are valid.

Multiplication of vectors by the matrix Σ^{-1} can be performed in $O(h^{-1}\log h^{-1})$ arithmetic operations using the Fast Fourier Transformatin [2]. Now let us consider the cross point case, i.e. the set S has cross points. To define the easily invertible matrix Σ we use the additive Schwarz method on the interface S. According to [3,4], partition the set S into parts in the following way. Let $p_i, \quad i = 1,2,\ldots,m_1$, be cross points of S (the cross points are points belonging simultaneously to boundaries at least of three subdomains $\Omega_i, \quad i = 1,2,\ldots,n$). Set $S_i = S \bigcap B(p_i, r_i), \quad i = 1,2,\ldots m_1$, where $B(p_i, r_i)$ is a ball of radius r_i with the center at p_i. We can choose r_i such that S_i will not contain any other cross points. The remaining part of S can be covered by the overlapping curvilinear open segments $S_i, \quad i = m_1+1,\ldots,m$, such that for any point $p \in S$

126

there exist $r > 0$ and S_i, where r is independent of h, such that $(B(p,r) \cap S) \subset S_i$. Define the subspaces V_i by $V_i := \left\{ \varphi^h \in V \mid \varphi^h(x) = 0, \quad x \overline{\in} S_i \right\}$.

Now, the following lemma is valid.

Lemma 3 *Let I^h be an uniform mesh on the segment $[-1,1]$ with the mesh size $h = \frac{1}{n}$. There exists a positive constant c_5 , independent of α and h , such that*

$$\|\tilde{\varphi}^h\|_{H^{\frac{1}{2}+\alpha}(I_h)} \leq c_5 \|\varphi^h\|_{H^{\frac{1}{2}+\alpha}(0,1)}, \quad \forall \varphi^h \in H_h(0,1),$$

for any constant $\alpha : |\alpha| < \frac{1}{2}$. Here the function $\tilde{\varphi}^h \in H_h(I^h)$ is defined in the following way:

$$\tilde{\varphi}^h(i \cdot h) = \varphi^h(i \cdot h),$$

$$\tilde{\varphi}^h(-i \cdot h) = (1 - i \cdot h)\varphi^h(i \cdot h), i = 0, 1, \ldots, n.$$

Using **Lemma 3**, it is easy to see that there exists a positive constant c_6 , independent of α and h , such that for any $\varphi^h \in V$ there exist $\varphi_i^h \in V_i$ such that the following is valid :

$$\left. \begin{aligned} &\varphi_1^h + \varphi_2^h + \ldots + \varphi_m^h = \varphi^h , \\ &c_6 \left(\|\varphi_1^h\|_{H(S)} + \ldots + \|\varphi_m^h\|_{H(S)} \right) \leq \|\varphi^h\|_{H(S)} \end{aligned} \right\} \tag{11}$$

According to [3,4,8] and (11), to define the matrix Σ, we can define the easily invertible norms for the subspaces V_i, $i = 1, 2, \ldots, m$. For $i = m_1 + 1, \ldots, m$ it can be done by (10). We will identify the matrices Σ_i from (10) and their extensions by zero outside S_i. Multiplication of vectors by the matrix Σ_i^+ can be performed in $O(h^{-1} \log h^{-1})$ arithmetic operations.

Construct now the matrix Σ_i for the space V_i for some $i = 1, 2, \ldots, m_1$. For the sake of simplicity, we omit the subscript i. Then S is a vicinity of the cross point p with k outcoming branches S_i. We can assume that each branch contains the following $r+1$ nodes $z_{0,1} = p, z_{1,i}, \ldots, z_{r,i}$,belonging to S. Assume that the branches are enumerated in accordance with the clockwise (or counter-clockwise) bypassing of the node p starting with the branch S_i. Let S_i is the branch between subdomain with the parameters α_{i-1} and α_i, $i = 1, 2, \ldots, k$, $\alpha_0 = \alpha_k$. Each function $\varphi^h \in V$ is in correspondence with the vector $\varphi \in R^{k \cdot r + 1}$ defined by $\varphi = (\varphi_0, \varphi_1, \ldots, \varphi_k)^T$, $\varphi_0 \in R$, $\varphi_i \in R^r$.

Here φ_0 is equal to the value φ^h at the point p and components of φ_i are equal to the values of the function φ^h at the corresponding nodes $z_{i,j}$ of S_i. Let us assume

$$\alpha_i \leq \alpha_k, \quad i = 1, 2, \ldots, k-1, \tag{12}$$

and decompose the space V into a vector sum of two subspaces

$$V = F_1 + F_2. \tag{13}$$

Here F_1 is the image of the extension operator T defined by :

$$F_1 = \left\{ \varphi \in V \mid \varphi = (\varphi_0, \varphi_1, \ldots, \varphi_1, \varphi_k)^T = T(\varphi_0, \varphi_1, \varphi_k)^T , \ \varphi_0 \in R, \varphi_1, \varphi_k \in R^r \right\},$$

and the subspace F_2 is the following :

$$F_2 = \left\{ \varphi \in V \mid \varphi = (0, 0, \varphi_2, \ldots, \varphi_{k-1}, 0)^T , \varphi_i \in R^r, \ i = 2, \ldots, k-1 \right\}.$$

According to the assumption (13) and [12], the following lemma is valid.

Lemma 4 *There exists a positive constant c_7, independent of α and h, such that for any $\varphi^h \in V$ there exist $\varphi_1^h \in F_1$ and $\varphi_2^h \in F_2$:*

$$\varphi_1^h + \varphi_2^h = \varphi^h , \quad \|\varphi_1^h\|_{H(S)} + \|\varphi_2^h\|_{H(S)} \le c_7 \|\varphi^h\|_{H(S)} .$$

The space F_1 has the simple structure and we can define a preconditioner for this space in the form :

$$\Sigma_1^+ = Q T \Sigma_0^{-1} T^* Q^* , \qquad \Sigma_0 = h \left(\frac{1}{h^2} A_0 \right)^{\frac{1}{2} + \alpha_k} ,$$

where the matrix A_0 of the order $2r + 1$ is defined in (10), the matrix Q is the corresponding permutation matrix.

The main reason of the decomposition (13) is the getting of zero value at the point p for functions from F_2. Then, if $\alpha_{i-1} \neq 0$ and $\alpha_i \neq 0$, we can define on S_i the norm, according to (10); cf. [12]. For the case $\alpha_i = 0$ the following lemma is valid.

Lemma 5 *Let $\alpha_i = \alpha_{i+1} = \ldots = \alpha_j = 0$ be valid.*
Then there exist positive constant c_8 and c_9, independent of α and h, such that for any $\varphi^h \in F_2$

a) if $\alpha_{i-1} > 0$ or $\alpha_{j+1} > 0$, then $c_8 \sum_{k=i}^{j+1} \|\varphi^h\|_{\tilde{H}^{\frac{1}{2}+\beta_k}(S_k)} \le \|\varphi^h\|_{\tilde{H}^{\frac{1}{2}+\alpha_{i-1}}(S_i)} +$

$$+ \sum_{k=i}^{j} \|\varphi^h\|_{H^{\frac{1}{2}}(S_k \cup S_{k+1})} + \|\varphi^h\|_{\tilde{H}^{\frac{1}{2}+\alpha_{j+1}}(S_j)} \le c_9 \sum_{k=i}^{j+1} \|\varphi^h\|_{\tilde{H}^{\frac{1}{2}+\beta_k}(S_k)},$$

b) if $\alpha_{i-1} < 0$ and $\alpha_{j+1} < 0$, then $c_8 \sum_{k=i}^{j+1} \|\varphi^h\|_{H^{\frac{1}{2}}(S_k \cup S_{k+1})} \le \|\varphi^h\|_{\tilde{H}^{\frac{1}{2}+\alpha_{i+1}}(S_i)} +$

$$+ \sum_{k=i}^{j} \|\varphi^h\|_{H^{\frac{1}{2}}(S_k \cup S_{k+1})} + \|\varphi^h\|_{\tilde{H}^{\frac{1}{2}+\alpha_{j+1}}(S_j)} \le c_9 \sum_{k=i}^{j+1} \|\varphi^h\|_{H^{\frac{1}{2}}(S_k \cup S_{k+1})} .$$

Using **Lemma 5**, we can decompose the space F_2 in the following way. Denote by J a set of subscripts

$$J = \{l_1, l_1 + 1, \ldots, r_1;\ l_2, l_2 + 1, \ldots, r_2;\ \ldots;\ l_m, l_m + 1, \ldots, r_m\},$$

such that corresponding groups of coefficients

$$\begin{aligned}
&\alpha_{l_1}, \quad \alpha_{l_1+1}, \quad \cdots, \quad \alpha_{r_1}; \\
&\alpha_{l_2}, \quad \alpha_{l_2+1}, \quad \cdots, \quad \alpha_{r_2}; \\
&\quad\quad \cdots\cdots\cdots\cdots\cdots \\
&\alpha_{l_m}, \quad \alpha_{l_m+1}, \quad \cdots, \quad \alpha_{r_m}
\end{aligned}$$

satisfy the condition *b)* of **Lemma 5**. Let us split S into an union of the following nonoverlapping parts :$\S = \bigcup_{i \in J} S_i + \bigcup_{j=1}^{m} S^{(j)}$ and $S^{(j)} = \bigcup_{p=l_j}^{r_j} S_p$, And now let us consider the corresponding decomposition of F_2:

$$F_2 = \sum_{i \in J} V_i + \sum_{j=1}^{m} V^{(j)}, \quad \text{where} \quad V_i = \left\{ \varphi^h \in F_2 \mid\ \varphi^h(x) = 0, \quad x \in S_i \right\}, \quad \text{and}$$

$$V^{(j)} = \left\{ \varphi^h \in F_2 \mid\ \varphi^h(x) = 0, \quad x \in S^{(j)} \right\}.$$

According to (10), we can define the easily invertible $\Sigma_{2,i}$ for $V_i, i \in J$, and according to [4,6,7], we can define the easily invertible $\Sigma_2^{(j)}$ for $V^{(j)}$, $j = 1, \ldots, m$. Identifying the matrices $\Sigma_{2,i}$ and $\Sigma_2^{(j)}$ and their extensions by zero outside S_i and $S^{(j)}$, we may define the global preconditioner for the vicinity of cross point

$$\Sigma^{-1} = \Sigma_1^+ + \sum_{i \in J} (\Sigma_{2,i})^+ + \sum_{j=1}^{m} (\Sigma_2^{(j)})^+.$$

That completes the construction of the preconditioner on the boundaries of the subdomains $\Omega_i, \quad i = 1, 2, \ldots, n$.

References

[1] Lions, P.L.: On the Schwarz alternating method. *1st International Symposium on Domain Decomposition Methods for Partial Differential Equations*, Glowinski, P.; Golub, G.H.; Meurant, G.; Periaux, S.; (eds.), SIAM, Philadelphia, PA, 1988.

[2] Marchuk, G.I.: Methods of Numerical Mathematics. *Springer, NY*, 1982.

[3] Matsokin, A.M.; Nepomnyashchikh, S.V.: Schwarz alternating method in subspaces. *Soviet Mathematics*, 29 (10), pp. 78-84, 1955.

[4] Matsokin, A.M.; Nepomnyashchikh, S.V.: Norms in the space of traces of mesh functions. *Sov. J. Numer. Anal. Math. Modeling*, Vol. 3, No. 3 , pp. 199-216, 1988.

[5] Matsokin, A.M.; Nepomnyashchikh, S.V.: Method of fictitous space and explicit extension operators. *Zh. Vychisl. Mat. Mat. Fiz*, 33, pp. 52-68, 1993.

[6] Nepomnyashchikh, S.V.: Domain decomposition and Schwarz methods in a subspace for the approximate solution of elliptic boundary value problems. *PhD thesis*, Computing Center of the Siberian Branch of the USSR Academy of Sciences, Novosibirsk, USSR, 1986.

[7] Nepomnyashchikh, S.V.: Domain decomposition method for elliptic problems with discontinuous coefficients. *4th Conference on Domain Decomposition methods for Partial Differential Equations*, Philadelphia, PA, SIAM, pp. 242-251, 1991.

[8] Nepomnyashchikh, S.V.: Method of splitting into subspaces for solving elliptic boundary value problems in complex-form domains. *Sov. J. Numer. Anal. Math. Modelling*, Vol. 6, Nr. 2, pp. 151-168, 1991.

[9] Nepomnyashchikh, S.V.: Mesh theorems on traces, normalization of function traces and their inversion. *Sov. J. Numer. Anal. Math. Modelling*, Vol. 6, Nr. 3, pp. 223-242, 1991.

[10] S.V. Nepomnyashchikh, Decomposition and fictitious domain methods for elliptic boundary value problems. *5th Conference on Domain Decomposition Methods for Partial Differential Equations*, Philadelphia, PA , SIAM, 1992.

[11] Nikol'skiĭ, S.M.: Approximation of functions of many variables and embedding theorems. (in Russian), Nauka, Moscow, 1977.

[12] Yakovlev, G.N.: On the traces of functions from spaces W_p^l on piecewise-smooth surfaces. (in Russian), *Mat Sbornik*, Vol. 74 (116), 4, pp. 526-543, 1967.

Sergeĭ V. Nepomnyashchikh
Computing Center
6, Lavrentier av.
Novosibirsk, 630090
Russia

Singularities in interface problems

Serge Nicaise

1 Introduction

It is well known that the variational solution of a second order elliptic boundary value problem on a polygonal domain of the plane (cf. [10, 12, 7, 4]) or the variational solution of an interface problem in the plane (cf. [8, 9, 1, 11, 6]) with data in $W^{k,p}$, where k is a nonnegative integer and $p > 1$, does not have the optimal regularity $W^{k+2,p}$, in general. This solution admits a decomposition into a regular part v in $W^{k+2,p}$ and into a singular part $\sum_i k_i S_i$, where the singular functions S_i depend only on the domain and on the considered boundary value problem and the coefficients k_i depend continuously on the data.

We have shown in [13, 5, 14, 15] that this type of decomposition can be obtained for interface problems on two-dimensional polygonal topological networks (roughly speaking, it is a network such that each face is a polygon). This is not surprising since this type of problems extends the boundary value problems mentioned above. Moreover, for numerical and mechanical applications, the knowledge of the exact formula for the coefficients of the singularities is of great importance. For elliptic boundary value problems on a domain of $\mathbf{R}^n$ with conical points, this was done in [2], while for boundary value problems on two-dimensional polygonal topological network, this was given in [5, 14, 15]. In this paper, for the sake of simplicity, we shall recall, on a simple interface problem, the results we obtained and refer to the above-mentioned papers for the exact proofs. In the sequel, we shall use the ordinary Sobolev spaces $H^m(\Omega)$ and $W^{m,p}(\Omega)$, defined in any book (see for instance [7, 15]) concerning boundary value problems.

2 The model problem

Let Ω be a simply connected open subset of $\mathbf{R}^2$ with a polygonal boundary (see Figure 1). We suppose that Ω is the union of two polygons Ω_1, Ω_2, so that the interface $\Gamma = \bar{\Omega}_1 \cap \bar{\Omega}_2$ is a linear segment. We also denote by Γ_i the remainder of the boundary of Ω_i, i.e., $\Gamma_i = \partial\Omega_i \setminus \Gamma$, $i = 1, 2$.

We consider the following boundary value problem: given $(f_1, f_2) \in L^2(\Omega_1) \times L^2(\Omega_2)$, we look for a solution $(u_1, u_2) \in H^2(\Omega_1) \times H^2(\Omega_2)$ of problem (2.1)-(2.4)

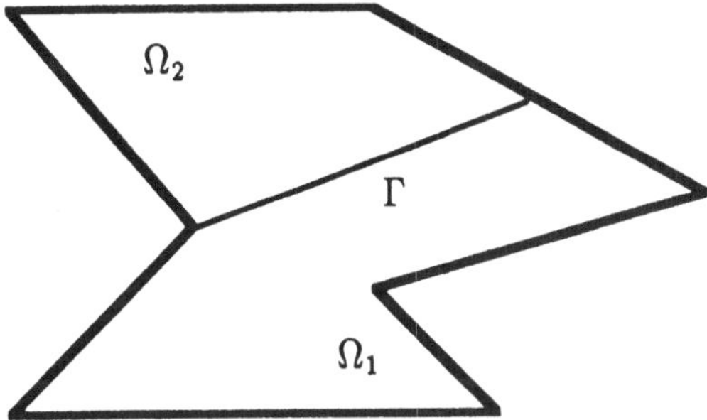

Figure 1: An interface problem

hereafter:

$$\Delta u_i = f_i \text{ in } \Omega_i, \forall i = 1, 2, \tag{2.1}$$

$$\gamma_i u_i = 0 \text{ on } \Gamma_i, \forall i = 1, 2, \tag{2.2}$$

$$\gamma u_1 = \gamma u_2 \text{ on } \Gamma, \tag{2.3}$$

$$\alpha_1 \gamma \frac{\partial u_1}{\partial \nu} - \alpha_2 \gamma \frac{\partial u_2}{\partial \nu} = 0 \text{ on } \Gamma, \tag{2.4}$$

where Δ denotes the Laplace operator in the plane, γ_i (resp. γ) is the trace operator on Γ_i (resp. Γ) and $\gamma\frac{\partial}{\partial \nu}$ represents the normal derivative along Γ (in a fixed direction). α_1, α_2 are two positive real numbers, assumed to be different (otherwise, if $\alpha_1 = \alpha_2$, then the previous problem (2.1)-(2.4) is a classical Dirichlet problem in Ω).

Remark 1 We can study more general problems: for intance condition (2.2) may be replaced by a Neumann boundary condition or a mixed boundary condition (Dirichlet condition on a part of the external boundary and Neumann one on the remainder). A first order tangential derivative may be added to the left-hand side of condition (2.4). As we said in the introduction, we can also study similar problems on two-dimensional polygonal topological networks (see Figure 2).

2.1 Weak solution

The first step in the study of problem (2.1)-(2.4) consists in proving existence of a weak solution. This means that we need the variational formulation of our problem: we set

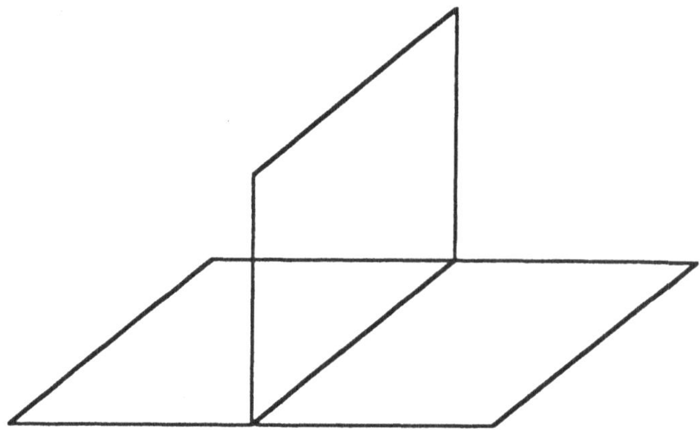

Figure 2: A two-dim. polygonal topological network

$$a : \overset{\circ}{H}{}^{1}(\Omega) \times \overset{\circ}{H}{}^{1}(\Omega) \longrightarrow \mathbf{C} : (u,v) \mapsto \sum_{i=1,2} \alpha_i \int_{\Omega_i} \nabla u \nabla \bar{v}\, dx.$$

This form is clearly well defined and continuous on $\overset{\circ}{H}{}^{1}(\Omega) \times \overset{\circ}{H}{}^{1}(\Omega)$. From Poincaré's inequality, we show that it is coercive, in other words, there exists a positive constant α such that

$$a(v,v) \geq \alpha \|v\|^2_{\overset{\circ}{H}{}^{1}(\Omega)}, \forall v \in \overset{\circ}{H}{}^{1}(\Omega).$$

This shows that the sesquilinear form a satisfies the assumptions of the Lax-Milgram lemma. Therefore for all $(f_1, f_2) \in L^2(\Omega_1) \times L^2(\Omega_2)$, there exists a unique solution $u \in \overset{\circ}{H}{}^{1}(\Omega)$ of

$$(2.5) \qquad a(u,v) = \sum_{i=1,2} \int_{\Omega_i} f_i \bar{v}\, dx, \forall v \in \overset{\circ}{H}{}^{1}(\Omega).$$

Now, we must clarify in what sense a solution of (2.5) satisfies (2.1)-(2.4). Denoting u_i the restriction of u to Ω_i, u clearly satisfy (2.2) and (2.3), due to the inclusion $u \in \overset{\circ}{H}{}^{1}(\Omega)$. Applying (2.5) with $v \in \mathcal{D}(\Omega_i)$, we obtain

$$\triangle u_i = f_i \text{ in the distributional sense}, \forall i = 1,2.$$

Consequently, owing to Theorem 1.5.3.10 of [7], $\gamma \frac{\partial u_i}{\partial \nu}$ is well defined as an element of $\tilde{H}^{1/2}(\Gamma)^*$. Moreover, from Theorem 1.5.3.11 of [7], the following

Green formula holds:

$$(2.6) \qquad a(u,v) = -\sum_{i=1,2} \alpha_i \int_{\Omega_i} \Delta u_i v \, dx + \langle \alpha_1 \gamma \frac{\partial u_1}{\partial \nu} - \alpha_2 \gamma \frac{\partial u_2}{\partial \nu} ; \gamma v \rangle,$$

for all $v \in \overset{\circ}{H}{}^1(\Omega) \cap W^{1,r}(\Omega)$, with $r > 2$. Comparing (2.6) with (2.5), we deduce that u satisfies (2.4).

In summary, we have proven the following statement:

Lemma 2 *For all $(f_1, f_2) \in L^2(\Omega_1) \times L^2(\Omega_2)$, there exists a unique solution $u \in \overset{\circ}{H}{}^1(\Omega)$ of (2.1) to (2.4).*

3   Regularity of the weak solution

Here we shall show that the variational solution of our problem admits a decomposition into a regular part and a singular one.

First of all, using the method of tangential differential quotients of Nirenberg, we show that u has the optimal regularity far from the corners. This means that

$$u_i \in H^2(\Omega_i \setminus V), \forall i = 1, 2,$$

for any neighbourhood V of the vertices of the Ω_i's.

It remains to study the behaviour of u near the vertices. For Dirichlet corners (i.e. for corners which are not endpoints of Γ, such as S_1 in Figure 3), the behaviour of u is wellknown from the results of Kondratiev [10], Maz'ya-Plamenskii [12], Grisvard [7], Dauge [4]. The behaviour of u near the vertex S (one a the endpoints of Γ, see Figure 3) was studied by Lemrabet [11] and Dobrowolski [6]. Our contribution is the expression for the coefficient of the singularities and the extension of all of that results to boundary value problems on two-dimensional polygonal topological networks.

Let us explain the asymptotics of u near S: we firstly need polar coordinates (r_i, θ_i) on Ω_i centred at S such that the half-line $\theta_i = 0$ contains the exterior side of Ω_i and the half-line $\theta_i = \omega_i$ contains the interface Γ (ω_i being the interior angle of Ω_i at S).

134

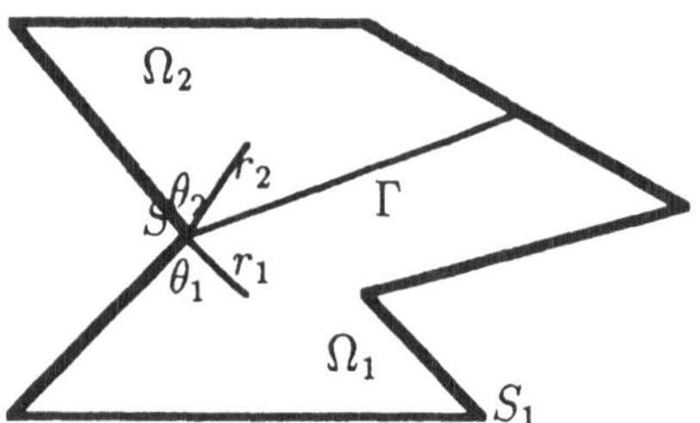

Figure 3: Polar coordinates

In order to define the singular functions of our problem, we introduce a vectorial operator Λ defined on $L^2(]0,\omega_1[) \times L^2(]0,\omega_2[)$ as follows:

$$D(\Lambda) = \{(v_1, v_2) \in H^2(]0,\omega_1[) \times H^2(]0,\omega_2[) \text{ satisfying}$$
$$v_1(0) = v_2(0) = 0,$$
$$v_1(\omega_1) = v_2(\omega_2),$$
$$\alpha_1 v_1'(\omega_1) + \alpha_2 v_2'(\omega_2) = 0\}.$$
$$\forall (v_1, v_2) \in D(\Lambda) : \Lambda(v_1, v_2) = (v_1'', v_2''),$$

where v_i' stands for the derivative of v_i.

This operator is a negative selfadjoint operator on $L^2(]0,\omega_1[) \times L^2(]0,\omega_2[)$ equipped with the inner product

$$((u_1, u_2), (v_1, v_2)) = \sum_{i=1,2} \alpha_i \int_0^{\omega_i} u_i(\theta_i)\bar{v}_i(\theta_i) \, d\theta_i.$$

Let us notice that α_1, α_2 are those used in (2.4).

The operator Λ has a discrete spectrum, which we denote by $\{-\lambda_n^2\}_{n\in\mathbf{N}}$, the corresponding orthonormalized eigenvector being denoted by $\{(v_1^n, v_2^n)\}_{n\in\mathbf{N}}$.

Remark 3 In this particular case, the λ_n's are the positive zeroes of

$$D(\lambda) = \alpha_1 \cos(\lambda\omega_1) \sin(\lambda\omega_2) + \alpha_2 \cos(\lambda\omega_2) \sin(\lambda\omega_1).$$

Each $-\lambda_n^2$ is of multiplicity one of associated eigenvector of the form

$$\left(C_1 \sin(\lambda_n\theta_1), C_2 \sin(\lambda_n\theta_2)\right),$$

where C_1, C_2 are two determined constants. Numerical examples of these λ_n's were given in [16].

Now we can state the following results:

Theorem 4 *Assume that $\lambda_n \neq 1$, for every $n \in \mathbf{N}$. Then for all $(f_1, f_2) \in L^2(\Omega_1) \times L^2(\Omega_2)$, the unique solution $u \in \overset{\circ}{H}^1(\Omega)$ of (2.1)-(2.4), admits the following expansion in a neighbourhood of the vertex S:*

$$(3.1) \qquad u_i = u_{i0} + \sum_{n:0<\lambda_n<1} k_n r_i^{\lambda_n} v_i^n(\theta_i),$$

for all $i = 1, 2$, where $u_{i0} \in H^2(\Omega_i)$ is the regular part of u_i. For every $\lambda_n \in]0,1[$, the coefficient k_n of the singularity $r^{\lambda_n} v^n$ is given by

$$(3.2) \qquad k_n = \sum_{i=1,2} \alpha_i \int_{\Omega_i} f_i \left(\Phi r_i^{-\lambda_n} v_i^n(\theta_i) - w_i^n\right) \, dx,$$

when Φ is a fixed cut-off function equal to 1 in a neighbourhood of S with a compact support disjoined to the other vertices and $w^n \in \overset{\circ}{H}^1(\Omega)$ is the unique solution (its existence following from the Lax-Milgram lemma) of

$$\triangle w_i^n = \triangle \left(\Phi r_i^{-\lambda_n} v_i^n(\theta_i)\right), \forall i = 1, 2,$$

satisfying (2.2)-(2.4).

The interest of formula (3.2) is that it does not need a numerical computation of the variational solution u of (2.1)-(2.4) to get a numerical approximation of the coefficient k_n. This is useful if we want to compute the coefficients for a great number of different data. In that case, we compute once and for all an approximation $w^{n,h}$ of w^n and the numerical approximation k_n^h of k_n is computed by (see [3] and the references cited there, for such considerations)

$$k_n^h = \sum_{i=1,2} \alpha_i \int_{\Omega_i} f_i \left(\Phi r_i^{-\lambda_n} v_i^n(\theta_i) - w_i^{n,h}\right) \, dx.$$

Let us notice that a non-exact formula (i.e. the solution u occurs in the formula) for k_n was given by Dobrowolski in [6].

Finally, let us remark that we may consider data f_i in $W^{m,p}(\Omega_i)$, for $m \in \mathbf{N}, p \in]0, +\infty[$. In that case, we get a similar decomposition into a regular part in $W^{m+2,p}$ and a singular part, with analogous formulas for the coefficients of the singularities.

References

[1] Ben M'Barek A., Mérigot M.: *Régularité de la solution d'un problème de transmission,* C. R. Acad. Sc. Paris, **280**, série A, 1975, 1591-1593.

[2] Bourlard M., Dauge M., Lubuma M.-S., Nicaise S.: *Coefficients des Singularités pour des problèmes aux limites elliptiques sur un Domaine à Points coniques I: Résultats généraux pour le problème de Dirichlet,* RAIRO Modél. Math. Anal. Numér., **24**, 1990, 27-52.

[3] Bourlard M., Dauge M., Lubuma M.-S., Nicaise S.: *Coefficients of the singularities for elliptic boundary value problems on domains with conical points III: Finite element methods on polygonal domains,* SIAM J. Numer. Anal., **29**, 1992, 136-155.

[4] Dauge M.: *Elliptic boundary value problems in corner domains. Smoothness and asymptotics of solutions,* L.N. in Math., **1341**, Springer Verlag, 1988.

[5] Dauge M., Nicaise S.: *Oblique derivative and interface problems on polygonal domains and networks,* Comm. in P.D.E., **14**, 1989, 1147-1192.

[6] Dobrowolski M.: *Numerical approximation of elliptic interface and corner problems,* Habilitationsschrift, Bonn, 1981.

[7] Grisvard P.: *Elliptic problems in nonsmooth Domains,* Monographs and Studies in Mathematics **21**, Pitman, Boston, 1985.

[8] Kellogg R. B.: *Singularities in interface problems,* Synspade 70, Ed. Hubbard, 351-400.

[9] Kellogg R. B.: *Higher order singularities for interface problems,* The Math. foundations of finite el. method with appl. to P. D. E., Ed. Aziz, Acad. Press, 1972, 589-602.

[10] Kondratiev V.A.: *Boundary value problems for elliptic equations in domains with conical or angular points,* Trans. Moscow Math. Soc., **16**, 1967, 227-313.

[11] Lemrabet K.: *Régularité de la solution d'un problème de transmission,* J. Math. Pures et Appl., **56**, 1977, 1-38.

[12] Maz'ya V. G., Plamenevskii B. A.: *Estimates in L^p and in Hölder classes and the Miranda-Agmon maximum principle for solutions of elliptic boundary value problems in domains with singular points on the boundary*, A.M.S. Trans. (2), **123**, 1984, 1-56.

[13] Nicaise S.: *Le laplacien sur les réseaux deux dimensionnels polygonaux topologiques*, J. Math. Pures et Appl., **67**, 1988, 93-113.

[14] Nicaise S.: *Polygonal interface problems - Higher regularity results*, Comm. in P.D.E., **15**, 1990, 1475-1508.

[15] Nicaise S.: *Polygonal interface problems*, Methoden und Verfahren der mathematischen Physik, **39**, Peter Lang, 1993.

[16] Nicaise S., Sändig A.M.: *General interface problems I and II*, Math. Meth. in the Appl. Sci., to appear.

Serge Nicaise
Université de Valenciennes et du Hainaut Cambrésis
L M^2 I and URA D 751 CNRS "GAT"
Institut des Sciences et Techniques de Valenciennes
Le Mont Houy, B.P. 311
F-59304 - Valenciennes Cedex (France)

Domain Control and Free Boundary Problems

Sergeĭ Okhezin and Valeriĭ Kalistratov [1]

Abstract

In this paper we consider a special type of optimization problems for elliptic, parabolic and hyperbolic partial differential equations, in which the space domains play the role of the controls. It is demonstrated that some of the free boundary problems may be formulated as shape optimization problems. To solve them, it is suggested to use the shape penalty method. It is also shown that the solution of one-dimensional and one-phase Stefan problems is a limit of the solutions of optimization problems for standard parabolic systems in a fixed domain. We prove theorems concerning the convergence of the approximate systems to the solution of the Stefan problem and present some numerical illustrations.

1   Introduction

There are many systems in which the control appears as the shape of the space domain where the considered process developes. In this paper we shall consider the explicit description of the mathematical models of such control problems for parabolic, elliptic and hyperbolic systems and apply this approach to free boundary problems. The general question we shall investigate is as follows [1]. Let Ω be a family $\{\Omega_\alpha \mid \alpha \in I\}$ of sets from R^n. For every $\alpha \in I$ the function $y\left(\cdot, \Omega_\alpha\right)$ denotes the solution of a state problem described by a partial differential equation defined in Ω_α.

Let $J : \Omega \times \{y\left(\cdot, \Omega_\alpha\right) \mid \alpha \in I\} \to R^1$ be a cost functional.

An abstract optimal shape design problem is stated as follows [1].

We want to find $\Omega^* \in \Omega$ such that

$$J\left(\Omega^*, y\left(\cdot, \Omega^*\right)\right) \leq J\left(\Omega_\alpha, y\left(\cdot, \Omega_\alpha\right)\right) \text{ for every } \alpha \in I.$$

Related to this, the following problems arise:

---

[1]This work was partially supported by Russian Foundation of Fundamental Researches grant 94-01-00231-a.

1. Existence of an optimal domain Ω^* .

2. Necessary optimality conditions.

3. Approximate calculation of the optimal domain.

4. Application of this approach to free boundary problems.

2   The shape penalty method

Let us consider the basic idea of the approach, which will be called a shape penalty method. Let $\Omega_i \subset R^n, i = 1, 2, \Omega_1 \subset\subset \Omega_2$ and

$$V(\Omega_1) = \{y \in L_2(\Omega_2) \mid y(x) = 0, \text{ a.e. } x \in \Omega_2 \setminus \Omega_1\}.$$

It should be noted that $V(\Omega_1)$ is a subspace of $L_2(\Omega_2)$. Let P be the metric projection of the space $L_2(\Omega_2)$ onto the subspace $V(\Omega_1)$

$$z - Pz = \chi z. \tag{1}$$

Here, $\chi(\cdot)$ is the characteristic function of the complement of Ω_1 with respect to Ω_2, that is

$$\chi(x) = \left\{ \begin{array}{l} 1, \text{ if } x \in \Omega_2 \setminus \Omega_1 \\ 0, \text{ if } x \in \Omega_1. \end{array} \right. \tag{2}$$

We shall also assume that the family $\{\Omega_\alpha \mid \alpha \in I\}$ has the following properties.

(i) For every $\alpha \in I$, Ω_α has the uniform C^m- regularity property (or strong local Lipschitz property) [2]. Here, m is a given positive integer.

(ii) There exists $\hat{\Omega}$, such that for every $\alpha \in I$

$$\Omega_\alpha \subset \hat{\Omega}.$$

(iii) A special type of set convergence is defined as follows :

$$\Omega_\alpha \to \Omega_\beta \text{ if } \text{Cl}\Omega_\alpha \to \text{Cl}\Omega_\beta$$

in the Hausdorff metric, when $\alpha \to \beta$ in R^1. $\text{Cl}A$ denotes the closure of a set A. It is assumed that the family $\{\Omega_\alpha \mid \alpha \in I\}$ is compact in the topology induced by this convergence.

Remark 2.1 It should be mentioned here that assumption (iii) must take place. This is due to the fact that, generally speaking, problem **1** has no solution without this property [3].

Next, we consider the shape penalty approximation.
Let A be a strongly elliptic operator of the type

$$A\left[\cdot\right] = -\sum_{i,j=1}^{n} \frac{\partial}{\partial x_i}\left(a_{ij}\left(x\right)\frac{\partial}{\partial x_j}\left[\cdot\right]\right) + a\left(x\right)\left[\cdot\right],\ x \in \hat{\Omega}. \tag{3}$$

1. For the original elliptic system

$$Ay = f \text{ in } \Omega_\alpha, \tag{4}$$

$$y = 0 \text{ in } \Gamma_\alpha = \partial\Omega_\alpha \tag{5}$$

the shape penalty approximation is

$$Ay_\epsilon + \frac{1}{\epsilon}\chi\left(x\right)y_\epsilon = f \text{ in } \hat{\Omega}, \tag{6}$$

$$y_\epsilon = 0 \text{ in } \hat{\Gamma} = \partial\hat{\Omega}. \tag{7}$$

2. For the original parabolic system

$$\frac{\partial y}{\partial t} + Ay = f \text{ in } Q_\alpha \subset \hat{Q} = (t_0, \theta) \times \hat{\Omega}, \tag{8}$$

$$y(t_0, x) = \varphi\left(x\right) \quad x \in \Omega_\alpha \tag{9}$$

$$y\left(t, x\right) = 0 \quad (t, x) \in \Sigma_\alpha = \prod_{t \in (t_0, \theta)} \partial Q_\alpha\left(t\right) \tag{10}$$

the shape penalty approximation is

$$\frac{\partial y_\epsilon}{\partial t} + Ay_\epsilon + \frac{1}{\epsilon}\chi\left(t, x\right)y_\epsilon = f \text{ in } \hat{Q}, \tag{11}$$

$$y_\epsilon\left(t_0, x\right) = \hat{\varphi}\left(x\right) = \begin{cases} \varphi\left(x\right), & x \in \Omega_\alpha \\ 0, & x \in \hat{\Omega} \setminus \Omega_\alpha, \end{cases} \tag{12}$$

$$y_\epsilon\,(t,x) = 0 \text{ in } \hat{\Sigma} = (t_0,\theta) \times \hat{\Gamma}. \tag{13}$$

Here,

$$\chi\,(t,x) = \begin{cases} 1, & (t,x) \in \hat{Q} \setminus Q_\alpha \\ 0, & (t,x) \in Q_\alpha. \end{cases} \tag{14}$$

3. For the original hyperbolic system

$$\frac{\partial^2 y}{\partial t^2} + Ay = f \text{ in } Q_\alpha \subset \hat{Q} = (t_0,\theta) \times \hat{\Omega}, \tag{15}$$

$$y\,(t_0,x) = \varphi\,(x) \quad x \in \Omega_\alpha, \tag{16}$$

$$\frac{\partial y}{\partial t}\,(t_0,x) = \psi\,(x) \quad x \in \Omega_\alpha, \tag{17}$$

$$y\,(t,x) = 0 \quad (t,x) \in \Sigma_\alpha \tag{18}$$

the shape penalty approximation is

$$\frac{\partial^2 y_\epsilon}{\partial t^2} + Ay_\epsilon + \frac{1}{\epsilon}\chi\,(t,x)\,\frac{\partial y_\epsilon}{\partial t} = f \text{ in } \hat{Q}, \tag{19}$$

$$y_\epsilon\,(t_0,x) = \hat{\varphi}\,(x) = \begin{cases} \varphi\,(x), & x \in \Omega_\alpha \\ 0, & x \in \hat{\Omega} \setminus \Omega_\alpha, \end{cases} \tag{20}$$

$$y_\epsilon\,(t_0,x) = \hat{\psi}\,(x) = \begin{cases} \psi\,(x), & x \in \Omega_\alpha \\ 0, & x \in \hat{\Omega} \setminus \Omega_\alpha. \end{cases} \tag{21}$$

The solutions of all systems formulated above are defined in generalized sense [4]. Our main results are stated in the following theorems [5].
For the elliptic system (4),(5) we introduce the notation

$$\hat{y}(x) = \begin{cases} y\,(x) \text{ - the general solution in } \Omega_\alpha \\ 0, \ x \in \hat{\Omega} \setminus \Omega_\alpha. \end{cases} \tag{22}$$

Theorem 2.1. *The convergence $y_\epsilon \to \hat{y}$ ($\epsilon \to 0$) in the weak topology of the Sobolev space $H_0^1\left(\hat{\Omega}\right)$ is uniform on the family of sets $\{\Omega_\alpha \mid \alpha \in I\}$.*

142

Analogously, we introduce for the parabolic system (8)-(10)

$$\tilde{y}(t,x) = \begin{cases} y(t,x) \text{ - the general solution in } Q_\alpha \\ 0, \quad (t,x) \in \tilde{Q} \setminus Q_\alpha. \end{cases}$$

We suppose that the family $\{Q_\alpha \mid \alpha \in I\}$ has the additional property

$$\forall (t,x), \; t \in [t_0, \theta], \; x \in \partial Q_\alpha(t) \Rightarrow \quad 0 \leq \cos\left(\vec{n}, \vec{t}\right) \leq \alpha_0 < 1.$$

Here, $\vec{n}$ is the inward normal unit vector, $\vec{t}$ is the time-oriented vector, α_0 is a given number.

Theorem 2.2. *The convergence $y_\epsilon \to \tilde{y}$ $(\epsilon \to 0)$ in the weak topology of the Sobolev space $H^1\left(\hat{Q}\right)$ is uniform on the family of sets Q_α.*

Remark 2.2. In the one-dimensional case, Theorem 2 may be proved without this additional property. For hyperbolic systems with constant coefficients we are able to prove an analogous version of Theorem 2 in the one-dimentional case, provided that at every instant of time the boundary of Q_α is contained in the corresponding sound cone. The uniform approximation enables us to apply this kind of shape approximation to solve the shape design problems for the systems (4),(5); (8)-(10); (15)-(18). The coefficient χ will play the role of the control function.

Remark 2.3. The following advantages of the shape penalty approximation method should be mentioned.

(i) The approximate systems are standard bilinear control systems defined in the fixed domain of integration.

(ii) The approximate systems may be solved by means of any standard numerical method, and suitable necessary and sufficient conditions of optimality can be obtained.

3 Application to the free boundary problems

The shape penalty approximation may be used for modelling a free boundary problem.

Let us consider for example the classical one-dimensional, one-phase Stefan problem.

It is well known [5] that the classical Stefan problem can be described by means of

(i) the state equation

$$\frac{\partial y}{\partial t}(t,x) = \frac{\partial^2 y}{\partial t^2}(t,x) \text{ in } Q(u) = \{(t,x) \mid 0 < t \le T, 0 < x < u(t)\},$$

$$(23)$$

(ii) the initial condition

$$y(0,x) = \varphi_0(x) \quad 0 \le x \le u_0, \tag{24}$$

(iii) the boundary conditions

$$y(t,0) = 0 \quad 0 < t \le T, \tag{25}$$

$$y(t,u(t)) = 0 \quad 0 < t \le T \tag{26}$$

(iv) and the Stefan condition on the moving boundary

$$\frac{\partial y}{\partial x}(t,u(t)) = -k \cdot \dot{u}(t) \quad 0 < t \le T, \tag{27}$$

$$u(0) = u_0. \tag{28}$$

The following shape design model for the Stefan problem [7,8] may be suggested. We describe the problem by

1. the family of the state equations ($\epsilon > 0$)

$$\frac{\partial y_\epsilon}{\partial t} - \frac{\partial^2 y}{\partial t^2} + \frac{1}{\epsilon}\chi_\epsilon(t,x;u(\cdot))\,y_\epsilon = 0$$

in the fixed cylindrical domain $D = (0,T) \times (0,X)$ with suitable $X > 0$,

2. the initial data

$$y_\epsilon(0,x) = \begin{cases} \varphi_0(x), & x \in [0,u_0] \\ 0, & x \in (u_0, X], \end{cases}$$

144

3. the boundary condition

$$y_\epsilon (t, x) = 0 \text{ in } \Sigma = [0, T] \times (\{0\} \cup \{X\}),$$

4. the minimization problem for the cost function

$$J (\epsilon, v (\cdot)) = \int_0^T \left(\frac{\partial y_\epsilon}{\partial x} (t, u (t)) + kv (t) \right)^2 dt \to \inf \text{ in } V [0, T]$$

5. and the set of admissible controls

$$V [0, T] = \{v (\cdot) \,|\, v (t) \in [0, V] \text{ a.e. } t \in [0, T]\},$$

$$V = k^{-1} \cdot \max \left\{ \left| \varphi_0 (x) (x - u_0)^{-1} \right| \;\middle|\; x \in [0, u_0] \right\}.$$

Here, $v (t) = \dot{u} (t)$, $u (0) = u_0$. The function χ_ϵ is a smooth regularization of the characteristic function χ introduced in the shape penalty method (14)

$$\chi_\epsilon = \begin{cases} 0, \; 0 \le x \le u (t) \\ \exp \left[- (x - u (t) - \epsilon)^2 (x - u (t))^{-2} \right], \; u (t) < x \le u (t) + \epsilon \\ 1, \; u(t) + \epsilon < x \le X. \end{cases}$$

The following theorems may be proved [9].

Theorem 3.1. *The functional $J (\epsilon, v (\cdot))$ is lower semicontinuous in the weak topology of $L_2 (0, T)$ and problem **4** has a solution.*

Theorem 3.2. *If $u_\epsilon \to u_*$ in $C [0, T]$ $(\epsilon \to 0)$, then*

$$\frac{\partial y_\epsilon}{\partial x} (\cdot, u_\epsilon (\cdot); u_\epsilon (\cdot)) \to \frac{\partial y_*}{\partial x} (\cdot, u_* (\cdot); u_* (\cdot)) \text{ in } L_2 (0, T).$$

Here, $y_* (t, x; u_* (\cdot))$ is the solution of the equation (22) in $Q (u_*)$.

Remark 3.1. This theorem is the counterpart of Theorem 2.2 on the uniform approximation by means of the shape penalty method.

Theorem 3.3. *If $u_\epsilon (\cdot)$, $y_\epsilon (\cdot, \cdot)$ is a solution of the minimization problem **4** and $u_0 (\cdot)$, $y_0 (\cdot, \cdot)$ is the solution of the Stefan problem, then*

$$u_\epsilon\left(\cdot\right) \to u_0\left(\cdot\right) \text{ in } C\left[0,T\right],$$

$$y_\epsilon\left(\cdot,\cdot\right) \to y_0\left(\cdot,\cdot\right) \text{ in } L_2\left(Q\left(u_0\right)\right), \quad \epsilon \to 0.$$

Remark 3.2. Theorem 3.3 permits to use the model **1-5** instead of the Stefan problem (i)-(iv) for a suitable parameter of approximation $\epsilon > 0$. The corresponding minimization problem may be solved by means of the gradient method.

The necessary conditions for the approximate system can be obtained in the following form.

If $v_\epsilon^*\left(\cdot\right)$ is a minimizing function for **4**, then

$$\forall v\left(\cdot\right) \in V\left[0,T\right] : \left\langle gradJ\left(\epsilon, v_\epsilon^*\left(\cdot\right)\right)\left(\cdot\right), v\left(\cdot\right) - v_\epsilon^*\left(\cdot\right)\right\rangle_{L_2(0,T)} \geq 0.$$

Here,

$$gradJ\left(\epsilon, v_\epsilon^*\left(\cdot\right)\right)\left(\tau\right) =$$

$$2k\left(\tfrac{\partial y_\epsilon}{\partial t} + kv_\epsilon^*\left(\tau\right)\right) + 2\int\limits_0^T \left(\tfrac{\partial y_\epsilon}{\partial x}\left(t, u_\epsilon^*\left(t\right)\right) + kv_\epsilon^*\left(t\right)\right) \times \tfrac{\partial^2 y_\epsilon}{\partial x^2}\left(t, u_\epsilon^*\left(t\right)\right) K\left(t,\tau\right) dt$$

$$-\epsilon^{-1} \int\limits_D y_\epsilon\left(t, x\right) w_\epsilon\left(t, x\right) \tfrac{\partial \chi_\epsilon}{\partial x}\left(t, x; u_\epsilon^*\left(\cdot\right)\right) K\left(t,\tau\right) dx dt,$$

$$K\left(t,\tau\right) = \left\{ \begin{array}{l} 0, \ \tau > t \\ 1, \ \tau \leq t, \end{array} \right.$$

$$\dot{u}_\epsilon^*\left(t\right) = v_\epsilon^*\left(t\right), \ u_\epsilon^*\left(0\right) = u_0.$$

The function w_ϵ is a generalized solution of the adjoint system in the space of distributions $D^*\left(0, T; \left(H_0^1\left(0, X\right) \cap H^2\left(0, X\right)\right)^*\right)$.

Adjoint system. The adjoint system is defined through the state equation

$$\tfrac{\partial w_\epsilon}{\partial t}\left(t, x\right) + \tfrac{\partial^2 w_\epsilon}{\partial x^2}\left(t, x\right) - \epsilon^{-1}\chi\left(t, x; u_\epsilon^*\left(\cdot\right)\right) w_\epsilon\left(t, x\right) =$$

$$2\left(\tfrac{\partial y_\epsilon}{\partial x}\left(t, u_\epsilon^*\left(t\right); u_\epsilon^*\left(\cdot\right)\right) + kv_\epsilon^*\left(t\right)\right) \delta'\left(x - u_\epsilon^*\left(t\right)\right) \quad \left(t, x\right) \in D,$$

146

the initial data

$$w_t\left(T,0\right)=0,\ 0\le x\le X$$

and the boundary conditions

$$w_\epsilon\left(t,0\right)=w_\epsilon\left(t,X\right)=0,\ 0\le t\le T.$$

Here, δ' stands for the derivative of the delta-function δ.

4 Numerical results

Some numerical calculations were performed. We used the projected gradient method to calculate the optimal control $v_\epsilon^*(.)$. The sweep method was used to solve the state and adjoint systems. The elements of the iteration sequence converging to the optimal control $v_\epsilon^*(.)$ are shown in Fig.1. The suitable solution $y_\epsilon\left(t,x;u_\epsilon^*(.)\right)$ is shown in Fig.2.

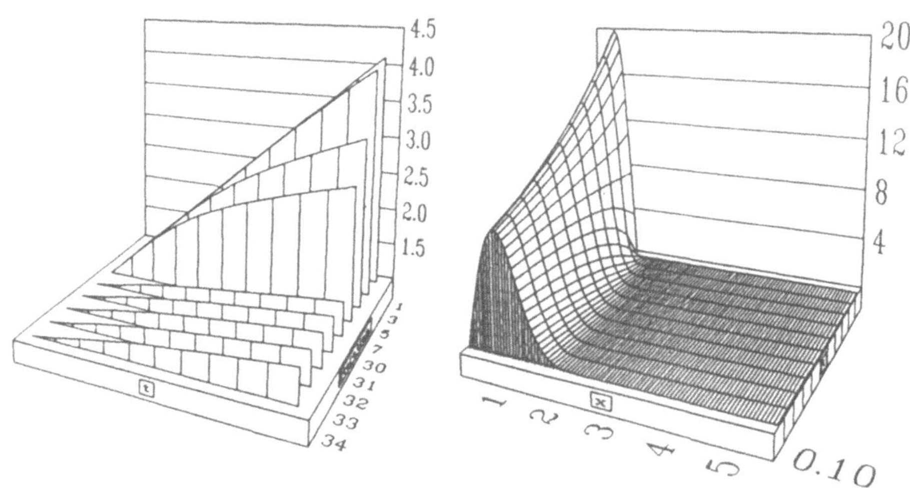

Fig.1   Fig.2

5 References

[1] Haslinger, J. and P. Neittaanmäki: Finite Element Approximation for Optimal Shape Design. Theory and Application. John Wiley & Sons Ltd. 1988.

[2] Adams, R.: Sobolev Spaces. Academic Press 1975.

[3] Osipov, Yu. and A. Suetov: The Existence of Optimal Shape for Elliptic Systems. Dirichlet Boundary Condition. Inst. Math. and Mech. Ural's Branch of Russian Academy of Sciences. Sverdlovsk 1990, 98 pp. (in Russian).

[4] Lions, J.-L.: Quelques Méthodes de Résolution des Problémes aux Limites Non Lineaires. Dunod, Paris 1969.

[5] Okhezin, S.P. On the Approximation in Domain Control Problem for Parabolic System. Prikl. Mat. Mekh. Vol. 54 (3) 1990, 361-365 (in Russian).

[6] Friedman, A.: Partial Differential Equations of Parabolic Type. Prentice Hall, Inc. 1964.

[7] Okhezin, S.P.: On the Mathematical Model for Stefan Problem. Diff. Uravnenia. Vol. 27 (6) 1991, 1042-1048 (in Russian).

[8] Okhezin, S.P.: Optimal Shape Design for Parabolic System and Two-Phase Stefan Problem. Intern. Series of Numer. Math. Vol. 106, 239-244, 1992. Birkhäuser Verlag Basel.

[9] Okhezin, S.P. and V.A. Kalistratov: The Optimization Model for Stefan Problem. Prikl. Mat. Meh. Vol.57 (3) 1993, 34-40 (in Russian).

Sergeĭ Okhezin and Valeriĭ Kalistratov
Department of Mathematical Physics
Ural State University
51, Lenin Ave.
Yekaterinburg 620083
Russia
e-mail: ohezins@urgu.e-burg.su

Wavelet Approximation Methods for Integral and Pseudodifferential Equations

Siegfried Prössdorf

Introduction

While initially typical applications of wavelets were concerned with signal and image analysis there have been recent attempts of applying wavelets to the solution of integral and differential equations (see e.g. [2], [34], [19], [21], [6], [33], [15]–[18], [28], [29]).

The present lecture which is essentially based on the papers [15]–[18] is concerned with generalized Petrov–Galerkin methods for elliptic periodic pseudodifferential equations in $I\!R^n$ covering classical Galerkin schemes, collocation, and quasiinterpolation. These methods are based on a general setting of multiresolution analysis, i.e., of sequences of nested spaces which are generated by refinable functions (see Sect. 1). In the first part of this lecture (Sects. 2 and 3) we develop a general stability and convergence theory for such a framework which recovers and extends many previously studied special cases. The key to the analysis is a local principle due to the author [30]. Its applicability relies here on a sufficiently general version of a so called discrete commutator property of wavelet bases. These results establish important prerequisites for developing and analysing in the second part (Sects. 4 and 5) methods for the fast solution of the resulting linear systems. These methods are based on compressing the stiffness matrices relative to wavelet bases for the given multiresolution analysis. Such a compression technique has been proposed in [2] where, however, only classical Galerkin methods and operators of order zero were discussed. It is shown (see [16]) that the order of the overall computational work which is needed to realize a certain accuracy is of the form $O(N(\log N)^b)$, where N is the number of unknowns and $b \geq 0$ is some real number. The theoretical results have been confirmed by numerical experiments (see Sect. 5).

At this stage we focus on the model case of periodic pseudodifferential equations to exploit the full advantages of Fourier transform techniques in connection with appropriate representations for the class of operators under consideration.

However, we do consider variable symbols and it should be mentioned that this class covers all the classical examples such as Hörmander's class, in particular, those arising in connection with boundary element methods. Moreover, most of the methods used here are of local nature and thus apply to the case of non–periodic equations, too.

1 Multiresolution and Wavelets

Multiresolution is by now a well–studied notion [24], [22], [3], [27]. Here we focus only on those variants which are useful for our purpose. More precisely, we will assume that $\varphi \in L_2(I\!R^n)$ is a fixed function with the following properties:

(P1) φ has compact support.

(P2) φ has (algebraic) linearly independent integer translates, by which we mean here that the mapping $c \to \sum_{k \in \mathbb{Z}^n} c_k \varphi(\cdot - k)$ is injective for *all* sequences c.

(P3) φ is *refinable*, i.e., there exists a finitely supported *mask* $\mathbf{a} = \{a_k\}_{k \in \mathbb{Z}^n}$ such that

$$\varphi = \sum_{k \in \mathbb{Z}^n} a_k \varphi(2 \cdot - k). \tag{1.1}$$

(P4) $\varphi \in C^{d'}(I\!R^n)$ for some $d' \in I\!N_0 = \{0, 1, 2, ...\}$.

(P5) φ is *accurate of degree d*, i.e., there exists $d \in I\!N_0$ such that for every polynomial p of degree $\deg(p) \le d$ one has

$$\sum_{k \in \mathbb{Z}^n} p(k)\varphi(\cdot - k) = p + q, \tag{1.2}$$

where q is some polynomial of degree $\deg(q) < \deg(p)$.

A few comments on these assumptions are in order. (P2) implies that the integer translates of φ are *stable*, i.e.,

$$\|c\|_{\ell_2(\mathbb{Z}^n)} \sim \left\| \sum_{k \in \mathbb{Z}^n} c_k \varphi(\cdot - k) \right\|_{L_2(\mathbf{R}^n)} \tag{1.3}$$

where $\|c\|^2_{\ell_2(I)} := \sum_{k \in I} |c_k|^2$. Moreover, (1.3) is known to imply

$$\sum_{k \in \mathbb{Z}^n} |\hat{\varphi}(x + k)|^2 > 0, \; x \in I\!R^n, \tag{1.4}$$

150

where for $f \in L_1(\mathbb{R}^n)$

$$\hat{f}(y) = \int_{\mathbb{R}^n} f(x) e^{-2\pi i y \cdot x} dx \,.$$

Furthermore, (P2) and (P3) imply $\hat{\varphi}(0) \neq 0$, so that one may assume without loss of generality that

$$\hat{\varphi}(0) = 1$$

which is the normalization assumed in (1.2). It is also known [5] that under the assumptions (P3) and (P4), property (P5) is guaranteed to hold for $d \geq d'$. But, of course, d could be larger than d' as shown by the B–spline or cube spline which are typical examples. Finally, it will be useful to know that (P3) and (P4) also imply the existence of some $\rho = \rho(a) \in (0,1)$ such that

$$|\partial_x^\alpha \varphi(x) - \partial_x^\alpha \varphi(y)| \leq \text{const } |x - y|^\rho \,, \quad |\alpha| = \alpha_1 + \cdots + \alpha_n = d' \,. \qquad (1.5)$$

Let $\langle \varphi \rangle^j$ denote the closure in $L_2(\mathbb{R}^n)$ of the span of the translates $\varphi(2^j \cdot -k)$, $k \in \mathbb{Z}^n$, so that by (1.1) $\langle \varphi \rangle^j \subset \langle \varphi \rangle^{j+1}$, $j \in \mathbb{Z}$. In order to generate any complement of $\langle \varphi \rangle^0$ in $\langle \varphi \rangle^1$ one needs the integer translates of $2^n - 1$ additional functions ψ_e, $e \in E_* = E \backslash \{0\}$, of the form

$$\psi_e = \sum_{k \in \mathbb{Z}^n} a_k^e \varphi(2 \cdot -k) \,, \qquad (1.6)$$

where $E = \{0,1\}^n$ represents $\mathbb{Z}^n / 2\mathbb{Z}^n$. It will be convenient to write sometimes $\varphi =: \psi_0$. The functions ψ_e are called *pre–wavelets* if

$$(\psi_e, \psi_{e'}(\cdot - k))_{\mathbb{R}^n} = 0 \,, \quad e, e' \in E, e \neq e', \ k \in \mathbb{Z}^n \,, \qquad (1.7)$$

such that

$$\langle \psi_e \rangle^j \perp \langle \psi_{e'} \rangle^\ell \,, \quad \ell > j \,, \quad e \in E, \ e' \in E_* \,. \qquad (1.8)$$

If in addition to (1.7) one also has $(\psi_e, \psi_{e'}(\cdot - k))_{\mathbb{R}^n} = \delta_{e,e'}\delta_{0,k}, e, e' \in E, k \in \mathbb{Z}^n$, the ψ_e for $e \in E_*$ are called *wavelets*. But full orthonormality in this sense, in conjunction with compact support, is hard to realize in a non–tensor–product setting. Examples of compactly supported pre–wavelets, however, are easier to obtain (see e.g., [7], [22], [31]).

In what follows, the ψ_e for $e \in E_*$ will be assumed to be either (pre–)wavelets or *biorthogonal wavelets*, i.e., there exist additional refinable functions $\tilde{\psi}_e, e \in E$, such that

$$(\psi_e, \tilde{\psi}_{e'}(\cdot - k))_{\mathbb{R}^n} = \delta_{e,e'}\delta_{0,k} \,, \quad e, e' \in E, \ k \in \mathbb{Z} \,. \qquad (1.9)$$

For examples of biorthogonal wavelets see [9], [10], [8]. Moreover, we will assume that the dual set $\tilde{\psi}_e, e \in E$, also satisfies (P1)–(P5) but possibly with respect to different parameters $d^{*'}, d^*$. In fact, it turns out to be very important to have the flexibility of choosing d^* larger than d. Roughly speaking, d determines the convergence rate of the solutions to the exact discrete problem, while d^* affects the compression rate, which has to be properly adjusted to achieve the desired overall accuracy for the solutions of the compressed scheme as well. This already indicates a certain limitation when working with (orthogonal) (pre–)wavelets. Since we wish to cover (pre–)wavelets and biorthogonal wavelets, we will always tacitly assume that $d = d^*$ whenever dealing with the former case. In the sequel, we will also assume that the masks $\mathbf{a}^e$ in (1.6) have finite support so that the ψ_e do have compact support.

In the following we need a *periodic* setting of multiresolution and wavelets. To this end, let $T^n = \mathbb{R}^n/\mathbb{Z}^n$ denote the n–dimensional torus. Let $\mathcal{L}_2 := \{g \in L_2(\mathbb{R}^n) : \sum_{k \in \mathbb{Z}^n} |g(\cdot - k)| \in L_2(T^n)\}$, so that for $g \in \mathcal{L}_2$ the periodization

$$[g] := \sum_{k \in \mathbb{Z}^n} g(\cdot + k)$$

as well as

$$g_k^j := 2^{nj/2}[g(2^j \cdot -k)]$$

are well–defined. Moreover, one can show that stability of $g \in \mathcal{L}_2$ in the sense of (1.3) implies (see [14], [28])

$$\|c\|_{\ell_2(\mathbb{Z}^{n,j})} \sim \left\| \sum_{k \in \mathbb{Z}^{n,j}} c_k g_k^j \right\|_{L_2(T^n)} \tag{1.10}$$

uniformly in $j \in \mathbb{N}_0$, where $\mathbb{Z}^{n,j} := \mathbb{Z}^n/2^j\mathbb{Z}^n$. Obviously, for φ as above, the spaces

$$V^j = \langle \varphi \rangle_0^j := \operatorname{span}\{\varphi_k^j : k \in \mathbb{Z}^{n,j}\} \tag{1.11}$$

satisfy

$$V^j \subset V^{j+1}, \quad \overline{\bigcup_{j \in \mathbb{N}_0} V^j} = L_2(T^n), \quad \bigcap_j V^j = \{0\}.$$

Moreover, since for $u \in L_2, g \in \mathcal{L}_2$,

$$([g], u) = (g, u)_{\mathbb{R}^n},$$

152

so that for $f, g \in \mathcal{L}_2$

$$([f], [g]) = (f, [g])_{\mathbb{R}^n} = ([f], g)_{\mathbb{R}^n},$$

the orthogonality relations (1.7) and (1.9) carry over to the periodic setting, so that, for instance, (1.9) reads

$$(\psi^j_{e,k}, \tilde{\psi}^{j'}_{e',k'}) = \delta_{e,e'}\delta_{j,j'}\delta_{k,k'}, \ e, e' \in E, j, j' \in \mathbb{N}_0, \ k \in \mathbb{Z}^{n,j}, \ k' \in \mathbb{Z}^{n,j'}.$$

Thus studying periodic multiresolution and wavelets is conveniently reduced to the classical setting.

In our numerical analysis the following direct and inverse estimates for the orthogonal projections $P_j : L_2(\mathcal{T}^n) \to V^j$ will be used frequently. Hereby $H^t = H^t(\mathcal{T}^n)$ denotes the classical (periodic) Sobolev space of (possibly real or negative) order t, equipped with the usual norm $\| \cdot \|_t$. In what follows, we will always assume that

$$-d - 1 \leq t < d' + \varrho, \ -d - 1 \leq t - r < d' + \varrho, \tag{1.12}$$

and τ will be a parameter ranging between t and $d + 1$.

Theorem 1.1 (see, e.g., [15]). *Under the above assumptions,*

$$\|u - P_j u\|_t \leq c2^{-j(\tau-t)}\|u\|_\tau, \ u \in H^\tau, \tag{1.13}$$

and

$$\|u\|_t \leq c2^{j(t-s)}\|u\|_s, \ u \in V^j, \ s \leq t, \tag{1.14}$$

hold uniformly in $j \in \mathbb{N}_0$, where the constant c is independent of u and j.
The following super–approximation property of the P_j, which is called *discrete commutator property*, is one of the main ingredients in our stability analysis.

Theorem 1.2 ([15]). *There exists $\delta \in (0,1)$ such that for any $f \in C^\infty(\mathcal{T}^n)$ and any $u \in V^j$, $s \leq \tau < d' + \varrho$,*

$$\|(I - P_j)(fu)\|_s \leq c2^{-j\delta}2^{-j(\tau-s)}\|f\|_{d+1,\infty}\|u\|_\tau,$$

where $\|f\|_{s,\infty} := \max_{|\gamma| \leq s} \sup_{x \in \mathcal{T}^n} |\partial^\gamma f(x)|$.
Notice that Theorem 1.2 implies, in particular, the convergence $(I - P_j)fP_j \to 0$, as $j \to \infty$, with respect to the operator norm.

2 Periodic Pseudodifferential Equations

To specify the type of equations, we begin with a simple example. Let Δ denote the n–dimensional Laplace operator. We seek one–periodic solutions of the Helmholtz equation

$$-\Delta u + \lambda u = f, \tag{2.1}$$

where $\lambda \geq 0$ and the right hand side f belongs to some Sobolev space H^t, $t \geq -1$. Defining the discrete Fourier transform by

$$\tilde{u}(k) := \int_{[0,1]^n} u(x)e^{-2\pi i x\cdot k}dx \,, \ k \in \mathbb{Z}^n \,,$$

where $x \cdot y = \sum_{j=1}^n x_j y_j$, we may rewrite (2.1) as

$$(\lambda - \Delta)u(x) = \sum_{k\in\mathbb{Z}^n} \tilde{u}(k)(\lambda + 4\pi^2|k|^2)e^{2\pi i k\cdot x} \,.$$

This is a special case of operators of the form

$$\sigma(x,D)u(x) := \sum_{k\in\mathbb{Z}^n} \sigma(x,k)\tilde{u}(k)e^{2\pi i k\cdot x} \,.$$

The function $\sigma \in C^\infty(\mathcal{T}^n \times \mathbb{Z}^n)$, which is called the *symbol* of the pseudodifferential operator $\sigma(x,D)$, is assumed to belong to a certain class $\sum^\mu$ for some $\mu \in \mathbb{C}$. Here $\sum^\mu$ is comprised of all symbols σ of the form

$$\sigma = \sigma_0 + \sigma_1 \,,$$

where $\sigma_0 \in C^\infty(\mathcal{T}^n \times (\mathbb{R}^n/\{0\}))$ is homogeneous of degree $\mu \in \mathbb{C}$, i.e.

$$\sigma_0(x,0) = 1 \,, \ \sigma_0(x,\lambda k) = \lambda^\mu \sigma_0(x,k) \,, \quad \text{for} \quad \lambda > 0 \,, \ k \neq 0 \,,$$

and for $r_1 < \text{Re } \mu = r$

$$|\partial_{(x)}^\beta \tau_{(k)}^\alpha \sigma_1(x,k)| \leq c_{\alpha,\beta}(1+|k|)^{r_1-|\alpha|} \quad \text{for} \quad x \in \mathcal{T}^n \,, \ k \in \mathbb{Z}^n \,.$$

Here

$$\tau = (\tau_1, ..., \tau_n)^T \,, \ (\tau_i f)(x) := f(x+e^i) - f(x) \,, \ i = 1, ..., n$$

and $e^i = (\delta_{ij})_{j=1}^n$. By $\mathcal{A}^r$ we denote the class of pseudodifferential operators of the form $A = \sigma(\cdot,D) + K$ where $\sigma \in \sum^\mu$, $r = \text{Re } \mu$ is the *order* of A, and $(Ku)(x) = \int_{\mathcal{T}^n} k(x,y)u(y)dy$ with $k \in C^\infty(\mathcal{T}^n \times \mathbb{R}^n)$ is a smoothing operator.

The results in [25], [26] show that all the classical pseudodifferential operators defined in [23] including those arising in connection with boundary integral equations [11], [13], [32], [35] are contained in $\mathcal{A}^r$. The operator $\sigma(x, D) \in \mathcal{A}^r$ is called *elliptic* if, for sufficiently large $|k|$, and $r = \operatorname{Re} \mu$,

$$|\sigma_0(x, k)| \geq c|k|^r, \ x \in \mathcal{T}^n.$$

Any $A \in \mathcal{A}^r$ is a bounded linear operator $A : H^t \to H^{t-r}, t \in \mathbb{R}$. This operator is Fredholm if and only if it is elliptic.

3 Generalized Petrov–Galerkin Schemes

The spaces V^j will be used as trial spaces for the approximate solution of

$$Au = f, \tag{3.1}$$

where $A \in \mathcal{A}^r$ and $f \in H^{t-r}$ is given. By Theorem 1.1, $V^j \subset H^t$ for each $t < d' + \rho$. In the following we will fix one such that t and assume that $\eta \in H^{r-t}(\mathbb{R}^n)$ is a fixed linear functional with support in some compact set $\Gamma \subset \mathbb{R}^n$.

Defining the functionals η_k^j by

$$\eta_k^j(f) := 2^{-nj/2}\eta(f(2^{-j}(\cdot + k))), \ k \in \mathbb{Z}^{n,j}, \tag{3.2}$$

we seek for an element $u^j \in V^j$ satisfying

$$\eta_k^j(Au^j) = \eta_k^j(f), \ k \in \mathbb{Z}^{n,j}. \tag{3.3}$$

Clearly, $\eta = \varphi$ corresponds to a classical Galerkin scheme, while $\eta = \delta(\cdot - x_0)$ give rise to collocation at the points $2^{-j}(k + x_0)$, $k \in \mathbb{Z}^{n,j}$, $j \in \mathbb{N}_0$. Other examples can be found in [15].

Our first objective is to study solvability of (3.3), and if this is the case, the convergence of the solution u^j as $j \to \infty$. The key to this problem is a suitable stability concept. To this end, let Q_j denote the projections corresponding to η_k^j (see [15, Sect. 4], for the precise definition of Q_j) such that (3.3) can be rewritten as operator equation

$$Q_j Au^j = Q_j f. \tag{3.4}$$

The scheme (3.3) or (3.4) is called (t, r)-*stable* if

$$\|Q_j Av\|_{t-r} \geq c\|v\|_t \tag{3.5}$$

for all $v \in V^j$ uniformly in $j \in I\!N_0$. Clearly, (3.5) means that the finite dimensional operators $A_j := Q_j A P_j : H^t \to H^{t-r}$ have uniformly bounded inverses $A_j^{-1} : H^{t-r} \to H^t$.

It turns out that the stability of (3.3) is equivalent to the ellipticity of a certain function λ which owns many features of the principal symbol σ_0 of the operator A and which is called the *numerical symbol* (more precisely, the "symbol of the Petrov–Galerkin scheme (3.3)"). The numerical symbol is defined by

$$\lambda(x,y) := \sum_{k \in \mathbb{Z}^n} \sigma_0(x,y+k)\hat{\varphi}(x+k)\overline{\hat{\eta}(y+k)} \tag{3.6}$$

for all $x,y \in T^n$ (provided that the series on the right hand side of (3.6) converges absolutely for all $x,y \in T^n$).

Theorem 3.1 ([15]). *The scheme (3.3) is (t,r)–stable if and only if the numerical symbol λ is elliptic, i.e.*

$$|\lambda(x,y)| \geq c|\theta(y)|^r$$

holds uniformly in $x \in T^n$, where

$$\theta = (\theta_1, ..., \theta_n)^T , \; \theta_j(y) = e^{2\pi i y_j} - 1 .$$

Example. Consider the nodal point collocation (i.e. $x_0 = 0$) with tensor product splines of degree d. Then $\hat{\eta} = 1$ and

$$\hat{\varphi}(\xi) = \prod_{l=1}^{n} \left(\frac{\sin \pi \xi_l}{\pi \xi_l} \right)^{d+1} .$$

Clearly, if d is odd, then λ is elliptic if and only if A is strongly elliptic, i.e. $\text{Re } \sigma_0 \geq \text{const} > 0$. Hence, in that case, Theorem 3.1 provides stability of the collocation with odd degree splines for strongly elliptic operators (cf. [1] for the one dimensional case and [15], [12] for the multidimensional case). Notice that, in the case of the classical Galerkin scheme, the ellipticity of the numerical symbol for strongly elliptic operators is a consequence of the stability property (1.4).

The proof of Theorem 3.1 proceeds in two steps (see [15]).

1. In the case when the principal symbol σ_0 does not depend on x, the corresponding stiffness matrix to (3.3) turns out to be a circulant with the eigenvalues $\lambda_k = 2^{jr}\lambda(2^{-j}k), \; k \in \mathbb{Z}^{n,j}, \; k \neq 0, \; j \in I\!N_0, \; \lambda_0 = 1$.

2. The general case can be reduced to the case 1 by means of the local principle [30].

156

By applying Theorem 1.1 in combination with usual Galerkin techniques one derives the following error estimates.

Theorem 3.2 ([15]). *Assume* (1.12) *and* $f \in H^{\tau-r}$ *for some* $\tau \geq t$, $d+1 \geq \tau-r$. *Suppose* (3.3) *is* (t,r)*-stable. Then*

$$\|u^* - u^j\|_t \leq c2^{-j(\tau-t)}\|u^*\|_\tau ,$$

where u^* *is the exact solution of* (3.1). *If in addition* $\eta \in H^{-t'}$ *for* $t' \leq t - r$, *one even has*

$$\|u^* - u^j\|_s \leq c2^{-j(\tau-s)}\|u^*\|_\tau \tag{3.7}$$

for $\max\{-d - 1, r\} \leq s \leq t$. *Finally, when* $Q_j = P_j$, (3.7) *holds for*

$$\max\{-d - 1, -d - 1 - r\} \leq s \leq \tau .$$

4 Fast Solution of Discrete Equations (Matrix Compression)

The basic strategy for compressing the stiffness matrices is to first decompose A_j into different components each of which is then to be compressed appropriately. We will consider two such decompositions, namely the *wavelet–decomposition*

$$A_j = Q_j A P_j = \sum_{\ell,\ell'=-1}^{j-1} (Q_{\ell+1} - Q_\ell)A(P_{\ell'+1} - P_{\ell'}) \tag{4.1}$$

where $Q_{-1} = P_{-1} = 0$, and the atomic decomposition

$$A_j = Q_j A P_j = \sum_{\ell=0}^{j}(Q_\ell A P_\ell - Q_{\ell-1} A P_{\ell-1}) . \tag{4.2}$$

When $Q_\ell = P_\ell$, i.e, for classical Galerkin schemes (4.1) would correspond to stiffness matrices relative to a wavelet basis. (4.2) is closely related to the atomic decomposition of Calderòn–Zygmund operators studied in [29]. An analogous compression scheme was also proposed in [2]. While there, however, only Galerkin schemes are considered, we can show [16] that it works actually for the much wider class of generalized Petrov–Galerkin methods described above. For the special case of collocation methods it is related to the multigrid approach studied in [4], [20]. The first step is to estimate the asymptotic behavior of the entries of A_j as j increases. To this end, it is convenient to introduce the following notation. We set

$$\mathcal{J}_\ell = \{I = (e, k, \ell) : e \in E_*, \ k \in \mathbb{Z}^{n,\ell}\}, \ \ell \geq 0, \ \mathcal{J}_{-1} = \{0\},$$

and

$$\psi_I = \psi_{e,k}^\ell \quad \text{for} \quad I = (e, k, \ell) \in \mathcal{J}_\ell.$$

It will be also convenient to put

$$|I| = 2^{-\ell}, \ I \in \mathcal{J}_\ell, \ \mathcal{J}^j = \cup_{\ell=-1}^j \mathcal{J}_\ell.$$

Theorem 4.1 ([16]). *Let $A \in \mathcal{A}^r$ and suppose that (P1)–(P5) and (1.12) hold. Then*

$$
\begin{aligned}
|\eta_{e,k}^j(A\psi_{e',k'}^j)| \ &+ \ \left|\eta_{e,k}^j(A\varphi_{k'}^j)\right| + \left|\eta_k^j(A\psi_{e,k'}^j)\right| \\
&\leq \ \frac{c2^{jr}}{(1 + 2^j|\Theta(2^{-j}(k - k'))|)^{n+d^*+1+r}} \,,
\end{aligned}
\tag{4.3}
$$

uniformly in $j \in \mathbb{N}$, $k, k' \in \mathbb{Z}^{n,j}$, $e, e' \in E_*$, where $c = \text{const} > 0$.

An estimate of the type (4.3) can also be found in [2] for the case $r = 0$ and for Daubechies wavelets.

The estimates (4.3) are stated so far only for the case that test functionals and trial functions have the same refinement level j. This is actually all that is needed for the atomic decomposition (4.2). To treat the decomposition (4.1) one has to interrelate *different levels*. We will present such estimates only for the case of classical Petrov–Galerkin schemes, i.e., when η is actually an L_2-function.

Theorem 4.2 ([16]). *Let $\frac{r}{2} < d^*$, $d' + \rho$, $d^* + 1 + \frac{r}{2} > 0$, and $n + r + d^* + 1 > 0$. Then for $A \in \mathcal{A}^r$ as above there exists $\delta \in (0, 1]$ such that*

$$\left|2^{-\ell'r/2}\eta_J(2^{-\ell r/2}A^\#\psi_I)\right| \leq \frac{c2^{-|\ell-\ell'|(\frac{n}{2}+\delta)}}{(1 + 2^{\min\{\ell,\ell'\}}|\Theta(2^{-\ell}k - 2^{-\ell'}k')|)^{n+1+d^*+r}} \tag{4.4}$$

uniformly in $I = (e, k, \ell) \in \mathcal{J}_\ell$, $J = (e', k', \ell') \in \mathcal{J}_{\ell'}$, where

$$A^\# := (I - Q_0)A(I - P_0). \tag{4.5}$$

Note that $A^\#$ annihilates V^0 which contains in this case only constants.

Note that the estimate (4.4) refers to the *diagonal scaling*

$$B^j := (b_{I,J})_{I,J \in \mathcal{J}^j}, \quad b_{I,J} = |I|^{\frac{r}{2}}|J|^{\frac{r}{2}}\eta_I(A\psi_J)$$

158

of A_j. The corresponding scaled matrix representation for A replaced by $A^{\#}$ is then given by

$$B^{\#,j} = \left(b^{\#}_{I,J}\right)_{I,J\in\mathcal{J}^j}, \qquad b^{\#}_{I,J} = \begin{cases} b_{I,J}, & I,J \in \mathcal{J}^j\backslash\mathcal{J}_{-1}, \\ 0, & \text{otherwise} \end{cases}$$

which means that the basis of V^0 has been discared from the whole basis of V^j.

We will now describe the type of matrix compression to be considered in the sequel. For certain sets $\mathcal{R} \subset \mathcal{J}^j \times \mathcal{J}^j$ and any $\mathbf{C} = (c_{I,J})_{I,J\in\mathcal{J}^j}$ we set

$$\mathbf{C}_{\mathcal{R}} := (c_{I,J}(\mathcal{R}))_{I,J\in\mathcal{J}^j}, \quad c_{I,J}(\mathcal{R}) := \begin{cases} 0, & (I,J) \in \mathcal{R}, \\ c_{I,J}, & \text{otherwise.} \end{cases}$$

We will continue using the notation $I = (e,k,\ell)$, $J = (e',k',\ell')$ for generic elements I, J of $\mathcal{J}^j$.

Theorem 4.3 ([16]). *Let $\frac{r}{2} < d^*$, $d' + \rho$, $d^* + 1 + \frac{r}{2} > 0$, $n + r + d^* + 1 > 0$. Let*

$$\mathcal{R}_{\epsilon_1} := \{(I,J) \in \mathcal{J}^j \times \mathcal{J}^j : \ell,\ell' \geq 0, 2^{\min\{\ell,\ell'\}}\left|\theta(2^{-\ell}k - 2^{-\ell'}k')\right| \geq \epsilon_1^{-1}\}. \quad (4.6)$$

Then

$$\left\| B^j - B^j_{\mathcal{R}_{\epsilon_1}} \right\| \leq c\min\{1, \epsilon_1^{d^*+1+r}\}$$

uniformly in $j \in \mathbb{N}$ and $\epsilon_1 > 0$. Moreover, for $\mathcal{R}_\epsilon := \mathcal{R}_{\epsilon_1} \cup \mathcal{R}_{\epsilon_2}$ with

$$\mathcal{R}_{\epsilon_2} := \{(I,J) \in \mathcal{J}^j \times \mathcal{J}^j : \ell,\ell' \geq 0, |\ell - \ell'| \geq \epsilon_2^{-1}\}$$

we have

$$\left\| B^{\#,j} - B^{\#,j}_{\mathcal{R}_\epsilon} \right\| \leq c\min\{1, \epsilon_1^{d^*+1+r}\} + 2^{-\delta/\epsilon_2}, \quad j \in \mathbb{N}, \epsilon_1, \epsilon_2 > 0,$$

where δ is some positive constant.

One can check that the first compression (4.6) leaves on the order of $N(\log_2 N)^2$ nonvanishing entries in $B^j_{\mathcal{R}_{\epsilon_1}}$, where $N = 2^{nj}$, while $B^{\#,j}_{\mathcal{R}}$ contains only $O(N)$ nontrivial entries.

The proof of Theorem 4.3 is based on Theorem 4.2 and a periodized version of the well–known Schur–lemma. Theorem 4.3 tells us which entries of the stiffness matrix B^j must be computed in order to guarantee a required accuracy. In principle, Theorem 4.3 provides a criterion which allows us to avoid the computation of the full stiffness matrix in the wavelet representation. A similar result has been obtained for the atomic decomposition (see [16]).

A more subtle question is to preserve, for the solution of the compressed system, the convergence rate given by Theorem 3.2. This has been realized in [16] by achieving a higher accuracy on lower scales. Such a compression technique needs a computational work of order $O(N(\log N)^b)$, where b is some positive number.

5 Some Remarks on Current Numerical Experiments

As a first attempt to support the preceding theory by numerical examples we consider the double layer potential equation over the boundary of the cube (of the tetrahedron, and of the bench, see [18]) although this does not quite fit the setting considered above. In fact, due to the lack of smoothness of the domain one expects the results to be worse in this case. At this stage we focus only on relating the compression rate to the accuracy of the corresponding approximate solution versus the solution for the uncompressed wavelet decomposition. The starting grid is chosen to be a triangulation of the boundary of the cube consisting of twelve congruent triangles, two for each face. Successively refined grids are obtained by dividing each triangle canonically into four congruent subtriangles. This results in a triangulation which is uniform except near the corners of the cube.

We employ a fully discretized collocation method based on piecewise linear Courant elements as trial functions. The collocation points are the vertices. Due to the simple structure of the trial functions we are content at this stage with approximating the entries of the stiffness matrix by a simple composite quadrature rule in combination with singularity subtraction. Here the integral over each triangle is replaced by a three point quadrature rule, where the node points are the midpoints of the edges. To facilitate reliable comparisons we compute the full stiffness matrix relative to the nodal basis. The transformation which would carry this matrix over to a stiffness matrix relative to a wavelet–type basis requires only knowledge about the corresponding mask coefficients. These coefficients are determined here by using only discrete biorthogonality conditions. They are then used to apply a pyramid–type scheme for the transformation. Among those entries which should be considered according to the a–priori information provided by our theoretical estimates also those will be discarded that stay below a certain threshold. The threshold is chosen in such a way that the compression does not increase the relative ℓ_2–error (i.e. the error between the approximate solution on the given grid and the approximate solution on the finest grid). The following table shows some excerpts of the numerical tests for the case of the bench.

level	# coll. pts.	# entries $\neq 0$	compression rate	rel. ℓ_2-error	thresh.
1	58	1.840	1.8	$4.4 \cdot 10^{-2}$	$3 \cdot 10^{-3}$
2	226	15.885	3.2	$1.8 \cdot 10^{-2}$	$1 \cdot 10^{-3}$
3	898	101.954	7.9	$6.3 \cdot 10^{-3}$	$3 \cdot 10^{-4}$
4	3586	796.353	16.1	$2.0 \cdot 10^{-3}$	$1 \cdot 10^{-4}$

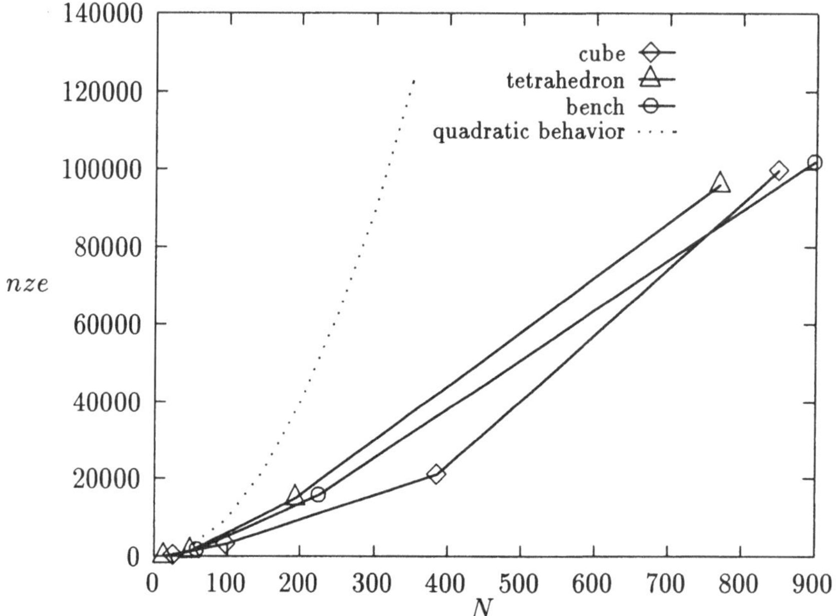

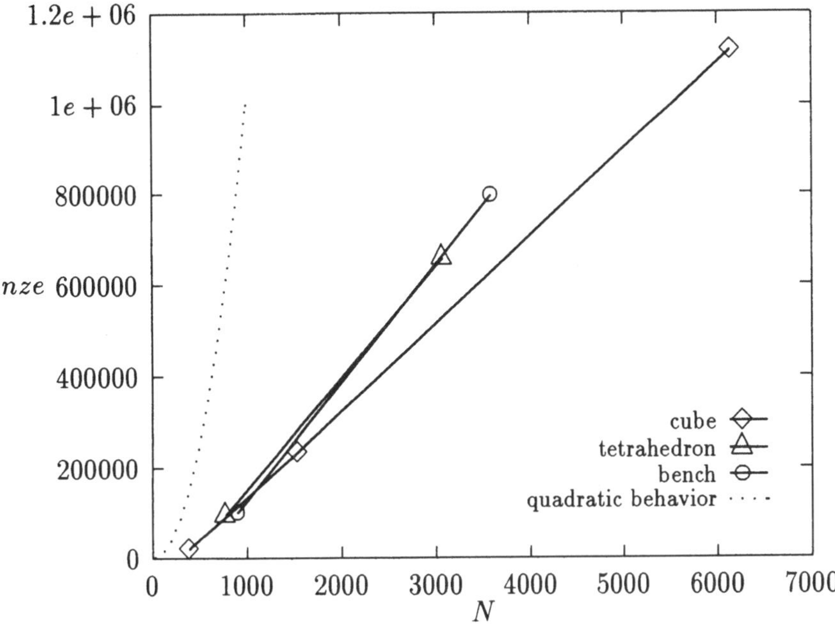

Figures 1 and 2: Number nze of nonzero elements of the transformed matrix after thresholding in dependence of N. Dotted line for comparison: quadratic behavior of the number of elements of the original stiffness matrix.

References

[1] Arnold, D.N., Wendland, W.L.: The convergence of spline collocation for strongly elliptic equations on curves. Numer. Math. **47** (1985), 317-431.

[2] Beylkin, G., Coifman, R., Rokhlin, V.: Fast wavelet transforms and numerical algorithms I. Comm. Pure and Appl. Math., Vol. XLIV (1991), 141–183.

[3] de Boor, C., DeVore, R., Ron, A.: On the construction of multivariate (pre)wavelets. Technical Summary Report No. 92–09, Center for the Mathematical Sciences, University of Wisconsin–Madison 1992.

[4] Brandt, A., Lubrecht, A.A.: Multilevel matrix multiplication and fast solution of integral equations. J. Comp. Phys. **90** (1991), 348–370.

[5] Cavaretta, A.S., Dahmen, W., Micchelli, C.A.: Stationary Subdivision. Memoirs of the American Math. Soc., Vol. 93, No. 453, 1991.

[6] Chui, C.K. (ed.): Wavelets: A Tutorial in Theory and Applications. Academic Press, Boston 1992.

[7] Chui, C.K., Stöckler, J., Ward, J.D.: Compactly supported box spline wavelets. Technical report, Preprint 1990.

[8] Cohen, A., Daubechies, I.: Non–separable bidimensional wavelet bases. Preprint AT & T Bell Laboratories 1992.

[9] Cohen, A., Daubechies, I., Feauveau, J.-C.: Biorthogonal bases of compactly supported wavelets. Comm. Pure and Appl. Math. **45** (1992), 485–560.

[10] Cohen, A., Schlenker, J.M.: Compactly supported bidimensional wavelet bases with hexagonal symmetry. Preprint AT & T Bell Laboratories 1992, to appear in Constructive Approximation.

[11] Costabel, M.: Boundary integral operators on Lipschitz–domains: Elementary results. SIAM J. Math. Anal. **19** (3) (1988), 613–626.

[12] Costabel, M., McLean, W.: Spline collocation for strongly elliptic equations on the torus. Numer. Math. **62** (1992), 511–538.

[13] Costabel, M., Wendland, W.: Strong ellipticity of boundary integral operators. J. Reine Angew. Math. **372** (1986), 39–63.

[14] Dahmen, W.: Locally finite decompositions of nested spaces and applications to operator equations. In: Algorithms for Approximation, M.G. Cox and J.C. Mason (eds), to appear.

[15] Dahmen, W., Prössdorf, S., Schneider, R.: Wavelet approximation methods for pseudodifferential equations I: Stability and convergence. Preprint No. 7, Institut für Angewandte Analysis und Stochastik, Berlin 1992; Math. Zeitschrift, to appear.

[16] Dahmen, W., Prössdorf, S., Schneider, R.: Wavelet approximation methods for pseudodifferential equations II: Matrix compression and fast solution. Advances in Computational Mathematics, 2nd Issue, 1 (1993), 259–335.

[17] Dahmen, W., Prössdorf, S., Schneider, R.: Multiscale methods for pseudodifferential equations. In: Recent Advances in Wavelet Analysis (eds. L.L. Schumaker and G. Webb), Academic Press 1993, 191–235.

[18] Dahmen, W., Kleemann, B., Prössdorf, S., Schneider, R.: A multiscale method for the double layer potential equation on a polyhedron. Preprint No. 76, Institut für Angewandte Analysis und Stochastik, Berlin 1993; Advances in Computational Mathematics (eds. H.P. Dikshit and C.A. Micchelli) 1994, to appear.

[19] Glowinski, R.R., Lawton, W.M., Ravachol, M., Tenenbaum, E.: Wavelet solution of linear and nonlinear elliptic, parabolic and hyperbolic problems in one space dimension. Preprint, Aware Inc., Cambridge, Mass. 1989.

[20] Harten, A., Yad–Shalom, I.: Fast multiresolution algorithms for matrix–vector multiplication. ICASE Report No. 92–55, October 1992.

[21] Jaffard, S.: Wavelet methods for fast resolution of elliptic problems. SIAM J. Numer. Anal. **29** (1992), 965–986.

[22] Jia, R.Q., Micchelli, C.A.: Using the refinement equation for the construction of pre–wavelets. In: Curves and Surfaces, P. Laurent, A. Le Méhauté, and L.L. Schumaker (eds.), Academic Press, New York 1991, 209–246.

[23] Kohn, J., Nirenberg, L.: On the algebra of pseudo–differential operators. Comm. Pure Appl. Math. **18** (1965), 269–305.

[24] Mallat, S.: Multiresolution approximation and wavelet orthogonal bases of $L^2(I\!\!R^n)$. Trans. Amer. Math. Soc. **315** (1989), 67–87.

[25] McLean, W.: Local and global description of periodic pseudodifferential operators. Math. Nachr. **150** (1991), 151–161.

[26] McLean, W.: Periodic pseudodifferential operators and periodic function spaces. Technical Report, University of New South Wales, Australia 1989.

[27] Meyer, Y.: Wavelets and Operators. Proc. Special Year in Modern Analysis, Urbana 1986/87.

[28] Meyer, Y.: Ondelettes et Opérateurs 1: Ondelettes. Hermann, Paris 1990.

[29] Meyer, Y.: Ondelettes et Opérateurs 2: Opérateur de Caldéron–Zygmund. Hermann, Paris 1990.

[30] Prössdorf, S.: Ein Lokalisierungsprinzip in der Theorie der Spline–
Approximationen und einige Anwendungen. Math. Nachr. **119** (1984),
239–255.

[31] Riemenschneider, S., Shen, Z.: Wavelets and pre–wavelets in low dimensions. Technical Report 1991.

[32] Schatz, A., Thomée, V., Wendland, W.: Mathematical Theory of Finite
and Boundary Element Methods. DMV Seminar, Birkhäuser Verlag, Basel,
Boston, Berlin 1990.

[33] Schumaker, L.L., Webb, G. (eds.): Recent Advances in Wavelet Analysis.
Academic Press, Boston 1993.

[34] Wavelet analysis and the numerical solutions of partial differential equations. Progress Report: June 1990, Aware Inc., Cambride, Mass.

[35] Wendland, W.L.: On some mathematical aspects of boundary element
methods. In: Mathematics of Finite Elements and
Applications V, J. Whiteman (ed.), Academic Press, London 1985, 193–
227.

Siegfried Prössdorf
Institut für Angewandte Analysis und Stochastik
Mohrenstraße 39
D – 10117 Berlin
e-mail: proessdorf@iaas-berlin.d400.de

On the Theoretical Foundation of CFD

Rolf Rannacher

Abstract

Computational Fluid Dynamics is of growing importance in science and technology. The idea is to simulate real fluids by approximately solving the Navier-Stokes equations on a computer. This paper addresses the related fundamental problem: When and why is it possible to compute fluid flow numerically with reliable control of the error? The results presented are based on joint work with C. Johnson, [6] and [7].

1 The Fundamental Problem in CFD

Computational Fluid Dynamics, or in short CFD, is a rapidly expanding industry which has now become of even greater importance in the study of flow phenomena than sole theory and experiment. This success is so obvious that there seems to be no need for further back-up concerning the principle question of whether real flows can be amenable to computations at all. By "flows", the mathematician, of course, means solutions of the underlying Navier-Stokes equations. At first glance, this question seems ridiculous, in view of the abundant number of research papers dealing with the various aspects of numerical methods in CFD and their practical applications. In fact, reviewing the existing literature particularly on the finite element method, one may gain the impression that most questions are more or less settled. There are all kinds of sophisticated discretization schemes and corresponding convergence proofs together with optimal-order error estimates. The problem caused by the incompressibility constraint is well understood today. Dominant convection is handled by upwinding or streamline diffusion. For the nonstationary case, robust time-stepping schemes have been designed and even the problem of turbulence is tackled by employing concepts from dynamical system theory. Finally, the range of computer simulations and the efficiency of the solution processes has enormously been improved by the use of new-generation supercomputers and the implementation of fast multi-level methods. But this enormous increase in computational power has also led to a new attitude concerning the reliability requirements for the numerical flow simulations. It used to be accepted, because of limited computing capacity, that a difficult problem was solved once by a single method on a single (usually too coarse!) mesh and that the results obtained were compared with experimental data which were normally considered to be more accurate (in representing solutions of the Navier-Stokes equations). Lately, however, the trend is shifting away from this "computational gambling" in the direction of actually "solving" the flow problem with controlled accuracy. This is reflected by the recent change in the editorial policy of some of the leading CFD journals the essence of which is that "the journals will not accept for publication any paper reporting the numerical solution of a fluid-dynamics problem that fails to address the task of systematic truncation error testing and accuracy estimation".

Unfortunately, the available theory in CFD provides only little help for this effort. The existing error estimates, although mathematically rigorous, are meaningless in most cases of interest as they involve constants which are left unspecified or which depend very unfavorably on the Reynolds number. So, there is a remarkable gap between theory and practice which needs to be closed. To this end, a new approch to error analysis in CFD has been developed in [6] and [7] which aims at reliable and efficient quantitative error control. It is based on a precise analysis of hydrodynamic stability coupled with the inherent power of Galerkin orthogonality. This paper discusses some of the concepts underlying this approach and its potential for serving as a rigorous basis in CFD.

2   A Critical Review of Theoretical CFD

As a model case, we consider the Navier-Stokes problem

Continuity equation: $\rho_t + \nabla\cdot(\rho\mathbf{u}) = 0$

Momentum equation: $(\rho\mathbf{u})_t + \nabla\cdot(\rho\mathbf{u}^T\mathbf{u}-\mu\nabla\mathbf{u}-\mu'\nabla\cdot\mathbf{u}\mathbf{I}) + \nabla p = \rho_0\beta(T-T_0)\mathbf{g} + \mathbf{f}$

Energy equation: $(\rho T)_t + \nabla\cdot(\rho T\mathbf{u} - \mu Pr^{-1}\nabla T) = 0$

Equation of state: $p \equiv F(\rho,T,\overline{\mathbf{u}})$

in the isothermal and incompressible limit, i.e., $\rho \equiv \rho_0$ and $T \equiv T_0$, together with appropriate initial and boundary conditions. For the numerical solution of this system, various methods of finite difference-, finite volume-, and finite element-(FE)-type have been developed (see, e.g., [2] and [8]). We strongly advocate the finite element Galerkin method which, due to its projection character, allows for a systematic *a-posteriori* error analysis as will be discussed below.

We consider any of the standard FE methods for solving the Navier-Stokes problem, characterized by a mesh size parameter h, together with an implicit time-stepping scheme with time step k (see [2], [3] and [11]). The available theoretical results on the convergence of this discretization typically have the form $\|u_h-u\| \leq Ch^r$ in the stationary case. The error constants C depend on the regularity and stability properties of the solution u, on the viscosity v, and in the non-stationary case in general also on the length of the time interval. The crucial question is: Of what size are the constants C? In the "normal" case of strongly elliptic (Laplace equation) or parabolic (heat equation) problems, one may expect that $C \approx 1\text{-}10^4$, depending on the particular situation. As a result, the error bound reduces by a factor 2^{-r} if the mesh size is halved which then, is hoped, also applies to the actual discretization error (at least for fine enough meshes). But the Navier-Stokes equations do not constitute a "normal" problem as they are of mixed elliptic-hyperbolic or parabolic-hyperbolic type involving degenerative effects in most interesting cases. If the characteristic length and velocity are normalized to one, the flow problem can be parametrized by the Reynolds number $Re=v^{-1}$, which for laminar flow may be assumed to be of size $Re \approx 1\text{-}10^5$. Then, the smallest spatial scale which has to be resolved is $h_{min} \approx \sqrt{v}$, and the length of the relevant time interval is $T \approx v^{-1}$ (see [4], [6]).

The dependence of the error estimates on the Reynolds number can be attributed to various sources:

1. In traditional error analysis, the nonlinearity is treated as a perturbation, i.e., the related error terms are absorbed, in the stationary case, into the (generally not dominant!) diffusive part $-v\Delta u$ (by using Young's inequality) which results in $C \approx v^{-1/2}$, and, in the nonstationary case, into the acceleration term u_t (by using Gronwall's inequality) which results in $C \approx \exp(cTv^{-1/2})$. Using a least-squares stabilization in the scheme, one can formally remove the additional v-dependence in the exponent to get $C \approx e^{cT}$ (see [5] and [10]).

2. In the presence of rigid boundaries (no-slip boundary condition), the regularity of the solution is essentially determined by the width of the boundary layer which is

168

$\delta \approx \nu^{1/2}$. Hence, $C \approx \max_\Omega |\nabla u| \approx \nu^{-1/2}$. This dependence on ν could in principle be suppressed by appropriately refining the mesh in the boundary layer.

3. The stability of the solution is normally introduced into the analysis by the linearization of the problem, e.g., in the stationary case, through the coercivity constant $\gamma(u)$ of the Fréchet derivative of the nonlinear operator taken at u , in the form $C \approx \gamma(u)$. The dependence of $\gamma(u)$ on ν can range between $\gamma(u) \approx \nu^{-1}$ and, in the worst case, $\gamma(u) \approx e^{c/\nu}$ (see the examples in [6] and [10]).

Since the stability of the flow to be computed is inherent to the given physical problem and cannot be improved by numerical tricks, the dependence of the error constant C on this property is crucial. It distinguishes "laminar" flows which are amenable to numerical computation from "turbulent" ones which in principle cannot be computed on their finest scales. For $Re \approx 1\text{-}10^5$, a dependence like $C \approx \nu^{-2}$ or even $C \approx e^{c/\nu}$ is not acceptable ($e^{20} > 10^8$ and $e^{100} > 10^{43}$!). Hence, to justify numerical computation for laminar flow, it is first necessary to identify appropriate stability concepts for *a-priori* and *a-posteriori* error estimation and, secondly, to specify cases where the relevant error constants are of "optimal" size, i.e., $C \approx \nu^{-1}$. We emphasize that only a sharp *a-posteriori* estimation of the error leads to a reliable and efficient adaptive control of the mesh-size.

3 A Realistic A-Posteriori Error Analysis

Let us first consider the stationary case. Using the uniform notation $(\cdot,\cdot)$ and $\|\cdot\|$ for the inner product and norm of $[L_2(\Omega)]^d$, for any $d \geq 1$, and setting $a(u,v) \equiv \nu(\nabla u, \nabla v)$, $b(u,v,w) \equiv (u \cdot \nabla v, w)$ and $A(u;v,w) \equiv a(u,v) + b(u,v,w)$, the variational formulation of the d-dimensional incompressible Navier-Stokes problem can be written in the compact form:

$$u \in V: \quad A(u;u,\varphi) = (f,\varphi) \quad \forall \; \varphi \in V \; . \tag{1}$$

The function space $V \equiv \{v \in H_0^1(\Omega)^d : \nabla \cdot v = 0\}$ contains the no-slip boundary condition as well as the incompressibility constraint. For simplicity, we consider a fully *conforming* FE discretization of (1) using subspaces $V_h \subset V$ of vector-functions which are piecewise polynomial with respect to certain decompositions T_h of Ω and exactly divergence-free (see,e.g., [2]). The discrete problem then reads:

$$\mathbf{u}_h \in V_h: \quad A(\mathbf{u}_h; \mathbf{u}_h, \varphi_h) = (\mathbf{f}, \varphi_h) \quad \forall \, \varphi_h \in V_h \; . \tag{2}$$

The error analysis for this scheme uses the Galerkin orthogonality property

$$A(\mathbf{u}; \mathbf{u}, \varphi_h) - A(\mathbf{u}_h; \mathbf{u}_h, \varphi_h) = 0 \quad \forall \, \varphi_h \in V_h \; , \tag{3}$$

and the derivative of the nonlinear form $A(\mathbf{u}; \mathbf{u}, \cdot)$ taken at the solution $\mathbf{u}$,

$$L(\mathbf{u}; \varphi, v) \equiv a(\varphi, v) + b(\mathbf{u}, \varphi, v) + b(\varphi, \mathbf{u}, v) \; .$$

In case of laminar flow, one may assume that the solution $\mathbf{u}$ is isolated, i.e., that the bilinear form $L(\mathbf{u}; \cdot, \cdot)$ is coercive on V, i.e.,

$$|v|_1 \leq C_1(\nu) \sup_{\varphi \in V, |\varphi|_1 = 1} L(\mathbf{u}; v, \varphi) \; , \tag{4}$$

where $C_1(\nu)$ is a constant depending on ν, and $|\cdot|_1 = \|\nabla \cdot\|$. In the "small data" case $|\mathbf{u}|_1 \approx \nu$, one has $C_1(\nu) \approx \nu^{-1}$. From (4), one concludes for the error $\mathbf{e} = \mathbf{u} - \mathbf{u}_h$ the estimate $|\mathbf{e}|_1 \leq c(\mathbf{u}) C_1(\nu) \inf_{\varphi_h \in V_h} |\mathbf{u} - \varphi_h|_1$, valid for small h. However, measuring the error of a flow simulation in the norm $|\cdot|_1$ appears unnatural as the (kinetic) energy is given in terms of the L_2-norm. Such L_2-error estimates are obtained in the finite element Galerkin context via so-called "duality arguments". For representing the error on the L_2-level, one uses the auxiliary "dual" problem

$$v \in V: \quad L(\mathbf{u}; \varphi, v) = (\varphi, \mathbf{e}) \quad \forall \, \varphi \in V \; . \tag{5}$$

Combining this with the orthogonality property (3), leads to an estimate of the form $\|\mathbf{e}\|_0 \leq c(\mathbf{u}) C_2(\nu) h^2$, for sufficiently small h, where the constant $C_2(\nu)$ is introduced by an assumed higher-order a-priori estimate for the dual problem,

$$\|v\|_2 \leq C_2(\nu) \sup_{\varphi \in V, \|\varphi\| = 1} L(\mathbf{u}; \varphi, v) \; , \tag{6}$$

where $\|\cdot\|_2$ denotes the H^2-norm.

So far the traditional a-priori error analysis. The error estimates obtained with unspecified constants $c(\mathbf{u})$ do not contain much practically useful information on the actual size of the error and particularly do not provide help for an adaptive control of the mesh-size. For this, one needs sharp a-posteriori error estimates involving only the approximate solution $\mathbf{u}_h$ and other computable quatities. Next, we will discuss the structure of such a-posteriori error estimates.

170

In the context of a Galerkin method the natural quantity entering an a-posteriori error estimate is the "residual" $R_h(u_h) \in V^*$ of the approximate solution defined by

$$\langle R(u_h), \varphi \rangle \equiv (f, \varphi) - A(u_h; u_h, \varphi) \qquad \forall\, \varphi \in V .$$

The evaluation of $R(u_h)$ in the natural norm of V^* is not very illustrative as it does not provide any information on the size of the error depending on the local mesh-size. We would prefer to measure it in an h-dependent L_2-norm. To this end, we consider a slightly modified bilinear form, which is related to the dual of the Fréchet derivative of $A(u; u, \cdot)$ taken "between" u and u_h,

$$L(u, u_h; \varphi, v) \equiv a(\varphi, v) + b(u, \varphi, v) + b(\varphi, u_h, v) .$$

With this notation and using the Galerkin orthogonality, there holds

$$L(u, u_h; e, v) = A(u; u, v) - A(u_h; u_h, v) = (R(u_h), v) = (R(u_h), v - v_h) .$$

From this we can then infer that

$$\|e\| \le c\, C_{2;h}(v) \|h^2 R(u_h)\| , \tag{7}$$

with a constant $C_{2;h}(v)$ which is given by the stability estimate

$$\|v\|_2 \le C_{2;h}(v) \sup_{\varphi \in V, \|\varphi\|=1} L(u, u_h; \varphi, v) .$$

In the above estimate, the mesh size h is considered as a piecewise constant function having on each element K the local value h_K , and c is a numerical constant independent of h, v and u . The residual norm in (7) can actually be evaluated as

$$\|h^2 R(u_h)\|^2 \approx \sum_{K \in T_h} \left\{ h_K^2 \|f + \nu \Delta u_h - u_h \cdot \nabla u_h - \nabla p_h\|_K^2 + h_K^{3/2} \|[\nu \partial_n u_h - n p_h]\|_{\partial K}^2 \right\} ,$$

where p_h is the discrete pressure associated with u_h , and $[\cdot]_{|\partial K}$ denotes the maximal jump across the boundary ∂K of the element K . On the basis of (7), the error control mechanism would then decrease the local mesh sizes h_K until the prescribed tolerance level TOL is reached:

$$C_{2;h}(v) \|h^2 R(u_h)\| \approx \text{TOL} .$$

This requires a reasonable upper bound for the stability constant $C_{2;h}(v)$. Clearly, this error control strategy is feasible only in cases with (at worst) $C_{2,h}(v) \approx v^{-1}$. A dependence like $C_{2,h}(v) \approx v^{-2}$ or worse would destroy the reliability in a numerical computation since the prescribed error tolerance would stay beyond reach with practical mesh-sizes. The v^{-1}-error growth seems to be generic for viscous flow problems and cannot be removed by modifying the discretization.

We now turn to the nonstationary case. Using the notation

$$(\mathbf{u},\mathbf{v})_I \equiv \int_I (\mathbf{u},\mathbf{v}) \, dt \, , \quad a_I(\mathbf{u},\mathbf{v}) \equiv \int_I a(\mathbf{u},\mathbf{v}) \, dt \, , \quad b_I(\mathbf{u},\mathbf{v},\mathbf{w}) \equiv \int_I b(\mathbf{u},\mathbf{v},\mathbf{w}) \, dt \, ,$$

$$A_I(\mathbf{u};\mathbf{v},\varphi) \equiv (\mathbf{v}_t,\varphi)_I + a_I(\mathbf{v},\varphi) + b_I(\mathbf{u},\mathbf{v},\varphi) + (\mathbf{v}^0,\varphi^0) \, , \quad F(\varphi) \equiv (\mathbf{f},\varphi)_I + (\mathbf{u}^0,\varphi^0) \, ,$$

on some time interval $I = [0,T]$ of length $T \approx v^{-1}$, the nonstationary Navier-Stokes problem may be written in compact form as

$$\mathbf{u} \in V(I): \quad A_I(\mathbf{u};\mathbf{u},\varphi) = F(\varphi) \quad \forall \, \varphi \in V(I) \, , \tag{8}$$

where $V(I) \equiv \{\mathbf{v} \in L^2(I;V), \mathbf{v}_t \in L^2(I;V^*), \mathbf{v}_{|t=0}=0\}$. This problem is approximated by the standard FE method in space combined with time discretization by the piecewise constant *discontinuous* Galerkin method. The latter is similar to the backward Euler scheme but uses time-averages rather than point values in the evaluation of the nonlinearity and the right-hand side. Let $0 = t_0 < t_1 < ... < t_{N+1} = T$, $I_n = [t_n,t_{n+1})$, $k_{|I_n} \equiv t_{n+1}-t_n$, $\mathbf{v}_+^n \equiv \lim_{s \to +0} \mathbf{v}(t_n+s)$, $\mathbf{v}_-^n \equiv \lim_{s \to -0} \mathbf{v}(t_n-s)$, $[\mathbf{v}^n] \equiv \mathbf{v}_+^n - \mathbf{v}_-^n$, $V_h(I) \equiv \{\mathbf{v}_h \in L^2(I;V_h): \mathbf{v}_{h|I_n} \equiv \text{const.}\}$, and

$$\tilde{A}_I(\mathbf{u};\mathbf{v},\varphi) \equiv \sum_{n=0}^{N} \{(\partial_t \mathbf{v},\varphi)_{I_n} + a_{I_n}(\mathbf{v},\varphi) + b_{I_n}(\mathbf{u},\mathbf{v},\varphi)\} + \sum_{n=0}^{N} ([\mathbf{v}^n],\varphi_+^n) + (\mathbf{v}_+^0,\varphi_+^0) \, .$$

Then, the space-time discretization reads

$$\mathbf{u}_h \in V_h(I): \quad \tilde{A}_I(\mathbf{u}_h;\mathbf{u}_h,\varphi_h) = F(\varphi_h) \quad \forall \, \varphi_h \in V_h(I) \, . \tag{9}$$

For this scheme we have an a-posteriori error estimate of the form

$$\max_I \|e\| \leq cL_h C_{2;h,k}(v) \{\max_I \|kd_t\mathbf{u}_h\| + \max_I \|kf\| + \max_I \|h^2R(\mathbf{u}_h)\|\} \, , \tag{10}$$

where $d_t\mathbf{u}_h \equiv k_n^{-1}(\mathbf{u}_{h,+}^n - \mathbf{u}_{h,-}^n)$ on I_n, $L_h = \max_{0 \leq n \leq N}(1+|\log k_n|)^{1/2}$, and the stability

172

constant $C_{2;h,k}(v)$ is defined by the a-priori estimate

$$\max_I \|v\| + |\log h_N|^{-1/2} \int_0^{T-h_N} \{\|v_t\|+\|v\|_2\} \, dt \ \leq \ C_{2;h,k}(v) \, \|v(\cdot,T)\| , \qquad (11)$$

for the dual of the linearized form (defined backward in time)

$$L_I(\mathbf{u},\mathbf{u}_h;v,\varphi) \ \equiv \ (-v_t,\varphi)_I + a_I(v,\varphi) + b_I(\mathbf{u},\varphi,v) + b_I(\varphi,\mathbf{u}_h,v) - (\varphi(T),v(T)) .$$

The evaluation of the stability constant $C_{2;h,k}(v,T)$ requires an accompanying sensitivity analysis for $\mathbf{u}_h$ backward in time.

4   A Remark on Hydrodynamic Stability

Traditional hydrodynamic stability theory is mostly "asymptotic linear stability theory" based on eigenvalue criteria. However, the theoretically obtained values for the "critical" Reynolds numbers at which instability and transition to turbulence should occur are often much larger than the experimentally observed values. For example, the theoretical critical Reynolds number of Poiseuille flow (parallel pipe flow) is about Re=5700 , but experiments show instability in a range of Re≈2000-40000. For Couette flow (parallel plane shear flow), the situation is even worse as theory predicts stability even for very high Reynolds numbers, but experiments show instability for rather moderate ones. This discrepancy was usually blamed on the deficiency of linearized stability theory which concerns only very small initial perturbations. But recent results show that this is not necessarily true (L.N. Trefethen, et al., [9]), i.e., linearized stability theory is perfectly correct but wrongly interpreted. In nonsymmetric eigenvalue problems, one has to care not only about critical eigenvalues but also about the size of the amplification factors related to positive but small eigenvalues. A model example of quantitative "instability" is given by the following simple ODE-system, with a small parameter $\varepsilon > 0$,

$$\dot{v}_1 + \varepsilon v_1 + v_2 = 0 , \quad \dot{v}_2 + \varepsilon v_2 = 0 ,$$

which has the solution $v_1(t) = e^{-t\varepsilon}v_1^0 - te^{-t\varepsilon}v_2^0$, $v_2(t) = e^{-t\varepsilon}v_2^0$. The 2-fold eigenvalue $\lambda = \varepsilon$ of the system matrix is positive, but small. This implies that a small initial perturbation $\|v^0\| \approx \varepsilon$ may grow over the time interval $[0,\varepsilon^{-1}]$ to size $\|v(\varepsilon^{-1})\| \approx 1$. This effect also occurs in nonlinear problems. But in this case, the final decay to zero, for $\varepsilon^{-1} \leq t \to \infty$, is practically irrelevant when the "nonlinear" instability has already taken over.

A flow problem which resembles the structure of the above model system is x_1-independent parallel pipe flow, i.e., flow in an infinitely long straight pipe $\Omega = \mathbf{R} \times \omega$ in the x_1-direction, where ω is the cross-section in the (x_2,x_3)-plane. The solution of this problem is $\mathbf{u} = (u_1,0,0)$, $u_1 = u_1(x_2,x_3,t)$, $p = p(t)$, satisfying

$$u_{1,t} - \nu\Delta u_1 = f_1, \quad \text{in } \omega\times I, \quad u_1 = 0 \quad \text{on } \partial\omega\times I,$$

where $\mathbf{f} = (f_1,0,0)$, $f_1 = f_1(x_2,x_3,t)$. Examples are flows in a long vertical pipe filled with fluid under gravity or in a long rotating pipe with variable rotation speed. The stability of this simple flow, assuming also x_1-independence of perturbations, can be analysed by elementary energy arguments yielding a constant of the form $C_2(\nu) \leq 2\nu^{-1}\max_{\omega\times I}|\nabla u_1|$ (see [6] and [7]). The active mechanism is again a decoupling of the v_2- and v_3-equations in the perturbation equation from the v_1-equation which allows an independent estimation of the (v_2,v_3)-component of the perturbation.

5   The Next Steps to Go

The elements of the presented a-posteriori error analysis have to be extended to more practical finite element schemes including streamline diffusion modifications and numerical integration. Further, the analytical stability approach should be applied to more interesting flows like Taylor-Couette flow, van Kármán vortex shedding and the Bérnard problem. This, however, seems to be a very difficult task in view of the scarce quantitative results in hydrodynamic stability theory.

The more important step is the estimation of the stability constants $C_{2;h}(\nu)$ and $C_{2;h,k}(\nu)$ in practically interesting situations. To this end, one would have to identify $L(\mathbf{u},\mathbf{u}_h;\cdot,\cdot) \approx L(\mathbf{u}_h,\mathbf{u}_h;\cdot,\cdot)$, and perform a kind of "backward" sensitivity analysis accompanying the "forward" computation of $\mathbf{u}_h$. First attempts in this direction for Poiseuille flow show that, indeed, the stability constant may be of rather moderate size, $C_2(\nu) \approx .02\nu^{-1}$ (see [1]). On the basis of this result, one should now convert the described concept of *a-posteriori* error estimation into a practical tool for adaptive error control. This program has to be left for future work.

References

[1]   Erksson, N., On the stability of pipe flow, Master of Sciences Thesis, Math. Dept., Chalmers University of Technology, Gothenburg, 1993.

[2]   Girault, V., Raviart, P.-A., Finite Element Methods for the Navier-Stokes Equations, Springer, Heidelberg, 1986.

[3]   Heywood, J.G., Rannacher, R., Finite element approximation of the nonstationary Navier-Stokes problem. Part 1, SIAM J.Numer.Anal. 19, 275-311 (1982), Part 2, ibidem, 23, 750-777 (1986), Part 3, ibidem, 25, 489-512 (1988), Part 4, ibidem, 27, 353-384 (1990).

[4]   Henshaw, W.D., Kreiss, H.O., Reyna, L.G., On the smallest scale for the incompressible Navier-Stokes equations, ICASE Report No. 88-8, 1988.

[5]   Johnson, C., The streamline diffusion finite element method for compressible and incompressible fluid flow, in "Finite Element Method in Fluids VII", Huntsville, 1989.

[6]   Johnson, C., Rannacher, R., Boman, M., Numerics and hydrodynamic stability: Towards error control in CFD, Preprint No. 93-12(SFB 359), IWR Universität Heidelberg, submitted to SIAM J.Numer.Anal.

[7]   Johnson, C., Rannacher, R., Error estimates for finite element methods for stationary nearly parallel pipe flow, Preprint SFB 359, Universität Heidelberg.

[8]   Peyret, R., Taylor, T.D., Computational Methods for Fluid Flow, Springer, Heidelberg, 1983.

[9]   Trefethen, L.N., Trefethen, A.,E., Reddy, S.C., Driscoll, T.A., A new direction in hydrodynamic stability: Beyond eigenvalues, Technical Report CTC92TR115 12/92, Cornell Theory Center, Cornell University, 1992.

[10] Tobiska, L., Verfürth, R., Analysis of a streamline diffusion finite element method for the Stokes and Navier-Stokes equations, to appear in SIAM J.Numer.Anal.

[11] Rannacher, R., On the numerical solution of the incompressible Navier-Stokes equations, Z.Angew.Math.Mech. 73, 203-216 (1993).

Rolf Rannacher
Universität Heidelberg
Institut für Angewandte Mathematik
INF 293, D-69120 Heidelberg, Germany
e-mail: rannacher@gaia.iwr.uni-heidelberg.de

Spectral Theory of Approximation Methods for Convolution Equations

Steffen Roch

Given a linear bounded operator A on a Banach space X, it is in general rather difficult to decide whether the equation $Ax = y$ with $x, y \in X$ is uniquely solvable for each right side y, and to solve this equation practically, or, in other words, to decide whether the operator A is invertible, and to determine its inverse A^{-1}. For that reason, besides this *"classical" invertibility* of A, other invertibility concepts are in discussion.

One of these concepts is that of *essential invertibility*. The operator A is essentially invertible (or a Fredholm, Noether, or Φ-operator) if its kernel has finite dimension (in contrast to $\mathrm{Ker}A = \{O\}$ in case of classical invertibility), and if its image possesses a finite dimensional complement in X (whereas $\mathrm{Im}A = X$ for classical invertibility). Clearly, it should be much simpler to study essential invertibility of an operator rather than its classical one. That this simplification is also of practical significance will be illustrated by the following example.

Let us start with recalling that both kinds of invertibility correspond to invertibility problems in associated Banach algebras. Indeed, by a theorem of Banach, an operator A is classically invertible if and only if it is invertible in the Banach algebra $L(X)$ of all linear and bounded operators on X. Further, A is essentially invertible if and only if it has an inverse modulo the ideal $K(X)$ of all linear compact operators on X or, equivalently, if the coset $A + K(X)$ is invertible in the Calkin algebra $L(X)/K(X)$. So we are led to the study of invertibility of elements in the Banach algebras $L(X)$ and $L(X)/K(X)$. In general, these algebras prove to be too large to tackle invertibility problems successfully, and therefore we have to restrict ourselves to look for inverses not in all of $L(X)$ or $L(X)/K(X)$ but in suitably chosen subalgebras. (This is a well known method: In order to determine, e.g., the inverse of the operator of the Sturm boundary value problem, one does not look for this inverse among all possible operators but only among the integral operators with continuous kernel function - the Green function of the operator.) Of course, here one has to take care to guarantee that invertibility in the larger algebra is equivalent to invertibility in the subalgebra.

Let now A be the operator $aI + bS$ where S is the singular integral operator on the real axis,

$$(Sf)(t) = \frac{1}{\pi i} \int_{-\infty}^{\infty} \frac{f(s)}{s-t}\, ds \quad , \quad t \in \mathbb{R},$$

and where the coefficients a and b are supposed to be continuous on $\mathbb{R} \cup \{\infty\} =:$ $\dot{\mathbb{R}}$ (continuity of a function b at ∞ means that both limits $\lim_{s\to+\infty} b(s)$ and $\lim_{s\to-\infty} b(s)$ exist and coincide). This operator is bounded on $L^2 = L^2(\mathbb{R})$. In place of the comprehensive algebras $L(L^2)$ and $L(L^2)/K(L^2)$ we consider their subalgebras $\mathcal{A}$, which is the smallest closed subalgebra of $L(L^2)$ containing all operators $aI + bS$ with a and b in $C(\dot{\mathbb{R}})$, and $\mathcal{A}/K(L^2)$.

The difficulties arising with classical invertibility find their algebraic expression in the fact that the algebra $\mathcal{A}$ is irreducible. Roughly speaking, this means that it is principally impossible to reduce invertibility in $\mathcal{A}$ to a simpler problem. On the other hand, one can show that the quotient algebra $\mathcal{A}/K(L^2)$ is even commutative!

The invertibility of elements of a commutative Banach algebra is a matter of Gelfand's spectral theory. This theory associates with each commutative Banach algebra a compact set T (the so-called maximal ideal space of the algebra), and with each element of this algebra a continuous complex-valued function on T (the so-called Gelfand transform of the element) with the property that the element is invertible if and only if this function does not vanish on T. In case of the algebra $\mathcal{A}/K(L^2)$ one finds that

$$T = \dot{\mathbb{R}} \times \{-1, 1\}\, ,$$

and the function corresponding to the coset $aI + bS + K(L^2)$ is just given by

$$(t, n) \mapsto a(t) + nb(t)\, . \tag{1}$$

Hence, the singular integral operator $aI + bS$ is essentially invertible if and only if the function (1) has no zero, and this condition is of course also necessary for the classical invertibility of $aI + bS$. But, in this special situation, there are additional conditions which, together with the essential invertibility, imply the classical one. More precisely, a corollary of a famous theorem of Coburn states that the singular integral operator $aI + bS$ is classically invertible if and only if it is essentially invertible and if the winding number of the curve $\{(a(t) + b(t))/(a(t) - b(t)), t \in \dot{\mathbb{R}}\}$ is zero. This shows, in particular, that essential invertibility is a *local* property (it only depends on the values of a

function), whereas classical invertibility is of *global* nature (the winding number depends on a function as a whole).

Now we are going to discuss the practical computation of the inverse of an operator $A \in L(X)$. Usually, A^{-1} is determined by a certain approximation method. Accordingly, the operator A is said to be *approximatively invertible* by a sequence (A_n) of approximation operators which converges strongly to A (i.e. $A_n x \to Ax$ as $n \to \infty$ for all $x \in X$) if and only if there is a number n_0 such that the equations $A_n x_n = y$ are uniquely solvable for all y and $n \geq n_0$, and if the sequence (x_n) of their solutions converges to a solution x of the equation $Ax = y$.

It is evident that this notion of invertibility is stronger than the classical one. As the latter, approximative invertibility is equivalent to invertibility in a certain Banach algebra again. In order to see this, recall that an operator A is approximatively invertible by a sequence (A_n) converging strongly to A if and only if the operator A is (classically) invertible and if the sequence (A_n) is stable. Stability of a sequence (A_n) means that there is an index n_0 such that the operators A_n are invertible for all $n \geq n_0$ and that $\sup_{n \geq n_0} \|A_n^{-1}\| < \infty$.

To get the desired formulation of stability in algebraic language, consider the collection $\mathcal{F}$ of all bounded sequences. On defining operations by $(A_n) + (B_n) := (A_n + B_n)$ and $(A_n)(B_n) := (A_n B_n)$ and a norm by $\|(A_n)\| := \sup_n \|A_n\|$, the set $\mathcal{F}$ even becomes a Banach algebra. Further, the set $\mathcal{G}$ of all sequences tending to zero in the norm forms a closed two-sided ideal in $\mathcal{F}$. Now, it is not hard to see that a bounded sequence (A_n) is stable if and only if its coset $(A_n) + \mathcal{G}$ is invertible in the quotient algebra $\mathcal{F}/\mathcal{G}$. (Indeed, if $(A_n) + \mathcal{G}$ is invertible then there are a bounded sequence (B_n) and sequences (C_n) and (D_n) tending to zero such that $A_n B_n = I + C_n$ and $B_n A_n = I + D_n$ whence, via a Neumann series argument, stability of (A_n) follows.)

The role of the algebra $\mathcal{F}/\mathcal{G}$ in approximative invertibility is the same as that of $L(X)$ in classical invertibility. Moreover, similarly to the classical invertibility, it turns out that the algebra $\mathcal{F}/\mathcal{G}$ is too large to work in successfully, and that its relevant subalgebras are irreducible again. The experiences from Fredholm theory suggest to look for an "*essential approximative invertibility*" which should allow to study first a simpler (possibly, local) problem, and then (under additional conditions) to turn back to proper approximative invertibility.

It was A. Kozak who first succeeded in developing a local theory of approximative invertibility (in case of the finite section method for two dimensional Toeplitz operators). But, as one can show, Kozak's approach is restricted to

178

rather special classes of operators (e.g. Toeplitz operators with continuous generating function). Later on, B. Silbermann proposed another approach with an essentially wider area of applications. He discovered a canonical procedure to construct ideals $\mathcal{J}$ of subalgebras of $\mathcal{F}$ with the following properties:

- essential approximative invertibility is invertibility modulo the ideal $\mathcal{J}$,

- invertibility modulo $\mathcal{J}$ is of local nature,

- essential approximative invertibility implies essential invertibility

and, in contrast to Fredholm theory, where Coburn's theorem appears as a lucky circumstance only,

- there are *well-defined* conditions which, together with essential approximative invertibility, imply approximative invertibility itself.

Let us illustrate the efficiency of this "lifting mechanism" by consideration of a simple spline Galerkin method for singular integral operators. We choose the spline space S_n of all piecewise constant functions over the partition $\frac{1}{n}\mathbb{Z}$ both as trial and test space, and we let L_n stand for the Galerkin projection of $L(L^2)$ onto S_n. In order to study the approximative invertibility of $aI + bS$ by the spline Galerkin method, we introduce the smallest closed subalgebra $\mathcal{B}$ of $\mathcal{F}$ which contains all approximation sequences $(L_n(aI + bS)|_{S_n})$ with $a, b \in C(\dot{\mathbb{R}})$. As already mentioned, both the algebra $\mathcal{B}$ and its quotient algebra $\mathcal{B}/\mathcal{G}$ are irreducible. Let now a subset $\mathcal{J}$ of $\mathcal{B}$ be defined by

$$\mathcal{J} = \{(L_n K|_{S_n}) + (C_n) \quad \text{with} \quad K \in K(L^2) \quad \text{and} \quad (C_n) \in \mathcal{G}\}.$$

This set (which is the analogue to the ideal $K(X)$ in Fredholm theory) owns just the properties we are interested in:

- *$\mathcal{J}$ is an ideal in $\mathcal{B}$.*

To verify that $\mathcal{J}$ is, e.g., a left sided ideal, let (A_n) be a sequence in $\mathcal{B}$ with strong limit A. Then

$$(A_n)(L_n K|_{S_n} + C_n) = ((A_n L_n - L_n A)K|_{S_n} + A_n C_n + L_n A K|_{S_n}),$$

and the sequence on the right hand side is in $\mathcal{J}$ since $A_n L_n - L_n A \to 0$ strongly as $n \to \infty$ which implies that the sequence $((A_n L_n - L_n A)K|_{S_n} + A_n C_n)$ is in $\mathcal{G}$, and since AK is compact again.

- *The algebra $\mathcal{B}/\mathcal{J}$ is commutative.*

This follows readily from the commutator relation (due to Arnold/Wendland and Prssdorf) $\|L_n fI - f L_n\| \to 0$ holding for each $f \in C(\dot{\mathbb{R}})$.

- *The ideal $\mathcal{J}$ can be lifted.*

Indeed, let, for example, the coset $(A_n) + \mathcal{J}$ be invertible from the right, that is, there are sequences $(B_n) \in \mathcal{B}$ and $(L_n K|_{S_n}) + (C_n) \in \mathcal{J}$ such that

$$A_n B_n = I + L_n K|_{S_n} + C_n.$$

Now suppose additionally the strong limit A of the sequence $(A_n L_n)$ to be invertible. Then set $B'_n = B_n - L_n A^{-1} K|_{S_n}$ to obtain

$$\begin{aligned}
A_n B'_n &= A_n B_n - A_n L_n A^{-1} K|_{S_n} \\
&= I + L_n K|_{S_n} + C_n - (A_n L_n - L_n A) A^{-1} K|_{S_n} - L_n K|_{S_n} \\
&= I + C'_n
\end{aligned}$$

with $(C'_n) = (C_n) - ((A_n L_n - L_n A) A^{-1} K|_{S_n}) \in \mathcal{G}$.

Thus, the sequence (A_n) is right sided invertible modulo $\mathcal{G}$. Similarly one verifies its invertibility from the left.

The Gelfand theory yields the following assertions now. The compact set associated with $\mathcal{B}/\mathcal{J}$ is

$$T = \dot{\mathbb{R}} \times [-1, 1] \,,$$

and the function on T corresponding to the coset $(L_n(aI + bS)|_{S_n})$ is given by

$$(t, z) \mapsto a(t) + z\, b(t). \tag{2}$$

Summarizing these results we get:

Theorem 1 *The Galerkin method applies to the operator $aI + bS$ with $a, b \in C(\dot{\mathbb{R}})$ if and only if this operator is invertible and if the function (2) does not vanish.*

Our next goal is the description of a more general (and more involved) situation. Let φ be any mother spline function, that is a bounded function with compact support which satisfies the conditions

$$\sum_{k \in \mathbb{Z}} \varphi(t - k) = 1 \quad \text{for all} \quad t \in \mathbb{R},$$

$$\sum_{k \in \mathbb{Z}} \int_{\mathbb{R}} \varphi(t+k)\,\overline{\varphi(t)}\,dt\, z^k \neq 0 \quad \text{for all} \quad z \in \mathbf{T}.$$

Set $\varphi_{kn}(t) = \varphi(nt - k)$, and let $S_n = S_n^{\varphi}$ stand for the smallest closed subspace of L^2 which contains all basis functions φ_{kn} with $k \in \mathbb{Z}$. In case φ is the characteristic function of the intervall $[0, 1)$, the space S_n^{φ} is just the spline space defined above and, moreover, all spaces of piecewise polynomials of degree d which are $d - 1$ times continuously differentiable are exactly of this form.

Let l^2 denote the Hilbert space of all complex valued sequences with norm $\|(x_k)_{k \in \mathbb{Z}}\| = (\sum_{k \in \mathbb{Z}} |x_k|^2)^{1/2}$. For each mother spline φ, the operators

$$E_n \;:\; l^2 \to S_n \quad , \quad (x_k) \mapsto \sum_{k \in \mathbb{Z}} x_k \varphi_{kn}$$

and

$$E_{-n} \;:\; S_n \to l^2 \quad , \quad \sum_{k \in \mathbb{Z}} x_k \varphi_{kn} \mapsto (x_k)$$

are correctly defined, they are linear and bounded, and $\sup_n \|E_n\|\,\|E_{-n}\| < \infty$.

Let furthermore $\mathcal{T}$ stand for the smallest closed subalgebra of $L(l^2)$ which contains the projection operator

$$P \;:\; l^2 \to l^2 \quad , \quad (\ldots, x_{-1}, x_0, x_1, \ldots) \mapsto (\ldots, 0, x_0, x_1, \ldots)$$

and all operators

$$T^0(a) \;:\; l^2 \to l^2 \quad , \quad (x_k) \mapsto (y_k) \quad \text{with} \quad y_k = \sum_{r \in \mathbb{Z}} a_{k-r} x_r$$

where a runs through the piecewise continuous functions on the unit circle $\mathbf{T}$, and a_r refers to the r th Fourier coefficient of a.

In place of the algebra $\mathcal{B}$ defined above, we now consider an essentially larger subalgebra, $\hat{\mathcal{B}}$, of $\mathcal{F}$, which contains all sequences of the form

$$(U_{-[tn]/n} E_n A E_{-n} U_{[tn]/n})$$

where $A \in \mathcal{T}$, $t \in \mathbb{R}$, $[r]$ denotes the integer part of the real number r, and U_s stands for the shift operator $(U_s f)(t) = f(t - s)$.

To get an impression of how large $\hat{\mathcal{B}}$ really is, let us mention that this algebra contains all sequences of the form $(L_n B|_{S_n})$ where B is a singular integral operator $aI + bS$ with a and b are allowed to have certain discontinuities, or B may be a Mellin convolution operator with continuous generating function, and where now L_n indicates a Galerkin (with, possibly, different trial and test spaces), a collocation, or a qualocation projection operator. Moreover, the approximation sequences resulting from certain composed quadrature rules for these operators B also belong to $\hat{\mathcal{B}}$.

If, finally, Y_t with $t \in \mathbf{T}$ denotes the operator

$$Y_t : l^2 \to l^2 \quad , \quad (x_k) \mapsto (t^{-k} x_k)$$

then one can show that, for each sequence $(A_n) \in \hat{\mathcal{B}}$, the strong limits

$$\text{s-lim } Y_t^{-1} E_{-n} A_n E_n Y_t E_n L_n =: W^t(A_n) \quad , \quad t \in \mathbf{T}$$

and

$$\text{s-lim } E_{-n} U_{-\left[\frac{sn}{n}\right]} A_n U_{\left[\frac{sn}{n}\right]} E_n =: W_s(A_n) \quad , \quad s \in \mathbb{R}$$

exist, and that the cosets

$$W_\infty(A_n) := E_{-n} A_n E_n + K(l^2)$$

are independent of n. Now one has the following general stability theorem.

Theorem 2 (2) *A sequence $(A_n) \in \hat{\mathcal{B}}$ is stable if and only if the operators (resp. the coset) $W^t(A_n)$ and $W_s(A_n)$ are invertible for all $t \in \mathbf{T}$ and $s \in \mathbb{R}$.*

One can prove this theorem in an analogous manner as Theorem 1, but the proof is essentially more extensive since the algebra $\hat{\mathcal{B}}$ is extremely non-commutative. Let us only mark the main steps of the proof.

First, one needs an ideal $\hat{\mathcal{J}}$ in $\hat{\mathcal{B}}$, which is essentially larger than the ideal $\mathcal{J}$ introduced above, to get a manageable quotient algebra $\hat{\mathcal{B}}/\hat{\mathcal{J}}$. This ideal can be defined by means of the following "lifting theorem" which goes, in its original form, back to Silbermann [3] and, in the present one, to Roch/Silbermann (see [1, 2, 4]).

Theorem 3 *Let $\mathbf{B}$ be a Banach algebra with identity and Ω be an arbitrary index set. Suppose that, fot all $t \in \Omega$, there are given a Banach space X^t, a unital homomorphism $W^t : \mathbf{B} \to L(X^t)$, and a closed two-sided ideal $\mathbf{J}^t$ of $\mathbf{B}$, such that the restriction of W^t onto $\mathbf{J}^t$ is an isomorphism between $\mathbf{J}^t$ and the ideal $K(X^t)$ of all compact operators on X^t. Let, further, $\mathbf{J}$ denote the smallest closed two-sided ideal of $\mathbf{B}$ which contains all ideals $\mathbf{J}^t$ with $t \in \Omega$. Then, an element b of $\mathbf{B}$ is invertible if and only if all operators $W^t(A_n)$ are invertible and if the coset $b + \mathbf{J}$ is invertible in $\mathbf{B}/\mathbf{J}$.*

In particular, now one needs a whole family of "lifting conditions" in order to turn over from the essential approximative invertibility modulo $\mathbf{J}$ to the proper approximative invertibility, viz. the invertibility of all operators $W^t(b)$.

In the modell case considered above one can take $\Omega = \{1\}$ and $W^1(A_n) =$ s-$\lim_{n\to\infty} A_n L_n$. In the general case we are going to discuss now, one identifies Ω with the unit circle $\mathbf{T}$ and defines ideals $\mathcal{J}^t$ by

$$\mathcal{J}^t = \{(E_n Y_t E_{-n} L_n^G K E_n Y_{t^{-1}} E_{-n}|_{S_n}) + \mathcal{G} \quad \text{with} \quad K \in K(L^2)\}$$

where L_n^G denotes the spline Galerkin projection with S_n both as trial and test space. The homomorphisms W^t involved by the lifting theorem are just the W^t defined above.

Let now $\mathcal{J}$ stand for the ideal which is generated by the $\mathcal{J}^t$ in accordance with Theorem 3. Even this ideal is not large enough to make the quotient algebra $\mathcal{B}/\mathcal{J}$ commutative! But, as one can show at least, this algebra possesses a non-trivial center (the center of an algebra consists of all elements which commute with each other element). More precisely, all cosets

$$(L_n f I|_{S_n}) + \mathcal{J} \tag{3}$$

with $f \in C(\dot{\mathbb{R}})$ lie in the center of $\mathcal{B}/\mathcal{J}$. In this situation, a generalization of the classical Gelfand theory works: the so-called *local principle by Allan*.

Theorem 4 *(Compare, e.g., [1, 4]) Let $\mathbf{B}$ be a Banach algebra with identity and $\mathbf{C}$ be a subalgebra of the center of $\mathbf{B}$ which contains the identity. For each s in the maximal ideal space T of the (commutative) Banach algebra $\mathbf{C}$, let $\mathbf{J}_s$ stand for the smallest closed two-sided ideal of $\mathbf{B}$ which contains all elements of $\mathbf{C}$ the Gelfand transform of which vanishes at s. Then an element b of $\mathbf{B}$ is invertible if and only if the cosets $b + \mathbf{J}_s$ are invertible in the quotient algebras $\mathbf{B}/\mathbf{J}_s$ for all $s \in \mathbf{T}$.*

In the context we are interested in, the algebras $\mathbf{B}$ and $\mathbf{C}$ from Theorem 4 correspond to the algebras $\mathcal{B}/\mathcal{J}$ and to its subalgebra generated by all sequences (3), respectively. The maximal ideal space of this central subalgebra can be identified with $\dot{\mathbb{R}}$ again, thus, Theorem 4 allows to localize the algebra $\mathcal{B}/\mathcal{J}$ over the points of the extended real axis.

Now, the second family of operators in Theorem 2 enters the scene. Namely, it turns out that the local coset $(A_n) + \mathcal{J} + \mathcal{J}_s$ is invertible in $(\mathcal{B}/\mathcal{J})/\mathcal{J}_s$ if and only if the operator $W_s(A_n)$ (resp. the coset $W_\infty(A_n)$ in case $s = \infty$) is invertible.

This is a consequence of the fact that the constant sequence $(W_s(A_n))$ and the sequence (A_n) itself show the same behavior locally at $s \in \dot{\mathbb{R}}$.

To finish the proof one has to check whether invertibility in $\mathcal{F}/\mathcal{G}$ (i.e. stability) is equivalent to invertibility in the restricted algebra $\mathcal{B}/\mathcal{G}$ (or, in other words, whether $\mathcal{B}/\mathcal{G}$ is *inverse closed* in $\mathcal{F}/\mathcal{G}$). For as long as we only consider operators on the Hilbert space L^2 this is no problem because all occuring algebras are C^*-algebras (and an element of a C^*-subalgebra of a C^*-algebra is invertible in the larger algebra if and only if it possesses an inverse in the subalgebra).

Let us finally mention some further developments.

- The general stability theorem (Theorem 2) holds on certain weighted Lebesgue spaces $L^p(w)$, too. The only new difficulty arising here is the inverse closedness of $\mathcal{B}/\mathcal{G}$ in $\mathcal{F}/\mathcal{G}$ since C^*-techniques do not work.

- One can define a version of the algebra $\mathcal{B}$ which contains spline approximation sequences for singular integral operators over arbitrary Lyapunov curves (having, possibly, corners, intersections, end points).

- There is a (technically expensive) generalization of Theorem 2 to the case of spline spaces which are generated by a finite number of mother splines (in contrast to the spaces introduced above which arise from only one mother spline). Examples of such spaces are that of all piecewise polynomials of degree d which are $d - e$ ($e \leq d$) times continuously differentiable. The generation of these spaces requires e mother splines.

- It is possible to include into the algebra $\mathcal{B}$ some cutting-off sequences and to derive corresponding stability conditions. This idea was brought forth by Chandler and Graham and then exploited by Elschner, Prssdorf and Rathsfeld in order to get better stability behavior by modifying the spline spaces in the neighborhoods of "singular points" (as corners or intersections of the underlying curves, or discontinuity points of the coefficient functions).

A detailed representation of the mentioned results can be found in [4].

Further generalizations of this approach were made to study approximation sequences for operators with Carleman shift or complex conjugation including the double layer potential operator (Didenko, Silbermann) and to multidimensional pseudodifferential operators (Prssdorf, Schneider for periodic spline spaces; Hagen, Roch, Silbermann for non-periodic spline spaces; Hoppe for singular

184

operators on the sphere; Elschner, Rathsfeld for potential operators on polyhedral domains). Finally (and rather surprisingly) the algebra $\mathcal{B}$ also contains sequences which are by no means related with splines, for example approximation sequences resulting from the finite section method for operators in the Toeplitz algebra $\mathcal{T}$ with respect to the standard basis of l^2 (Roch), and sequences coming from the finite section method for singular integral operators over $[-1, 1]$ with respect to weighted Chebyshev polynomials (Junghanns, Roch, Weber).

References

[1] Prößdorf, S., Silbermann, B.: Numerical analysis for integral and related operator equations. Akademie-Verlag, Berlin 1991, and Birkhäuser Verlag, Basel-Boston-Berlin 1991.

[2] Roch, S.: Nichtkommutative Gelfandtheorien und ihre Anwendung in Operatortheorie und numerischer Analysis. Habilitationsschrift, TU Chemnitz 1992.

[3] Silbermann, B.: Lokale Theorie des Reduktionsverfahrens für Toeplitzoperatoren. Math. Nachr. 104 (1981), 137-146.

[4] Hagen, R., Roch, S., Silbermann, B.: Spectral theory of approximation methods for convolution operators. Birkhäuser Verlag, Basel (to appear in 1994).

Steffen Roch
Technische Universität Chemnitz-Zwickau
Fakultät für Mathematik
PSF 964
D - 09009 Chemnitz
e-mail sroch@mathematik.tu-chemnitz.de

The Convergence of the Cascadic Conjugate-Gradient Method under a Deficient Regularity

Vladimir Shaĭdurov

1 Introduction

In this paper, we deal with a cascadic conjugate-gradient method (shortly called CCG-algorithm). This algorithm is a simpler kind of multigrid (multilevel) methods. We define it recurrently for discrete symmetric positive-definite problems on a sequence of grids. At the coarsest grid, the linear system is solved directly. At the finer grid, the system is solved iteratively by the conjugate-gradient method. A starting guess is an interpolation of the approximate solution from the previous grid. We do not implement any preconditioning or restriction onto a coarser grid. Nevertheless, CCG-algorithm has the same optimal property as the multigrid method. Namely, this algorithm converges with a rate which is independent of an amount of unknowns and a number of grids. In [6], this property was proved for a two-dimensional elliptic second-order Dirichlet problem in a convex polygon Ω, where a solution u belongs to space $H^2(\Omega)$. Here we study the problem in a non-convex polygon Ω, where $u \notin H^2(\Omega)$ due to a strong growth of second derivatives near some angular points. We prove the rate of convergence in two cases: for uniform grids and for grids with special refinement near some angular points. The paper [1] contains impressive numerical examples both for usual and deficient regularity.

2 The algebraic formulation of the cascadic algorithm

Assume that we have a sequence of finite-dimensional vector spaces

$$M_0, M_1, \ldots, M_l \tag{1}$$

186

equipped with inner products $(\,\cdot\,,\,\cdot\,)_i$, $i = 0, 1, \ldots, l$. Assume also that linear „interpolation" operators

$$I_i : M_i \to M_{i+1}, \quad i = 0, 1, \ldots, l-1, \tag{2}$$

and linear ivertible operators

$$L_i : M_i \to M_i, \quad i = 0, \ldots, l, \tag{3}$$

are given. Then, the main objective of the proposed algorithm consists in solving the following problem in a vector space M_l: *for a given $f_l \in M_l$, find $v_l \in M_l$ such that*

$$L_l v_l = f_l. \tag{4}$$

The main feature of the cascadic algorithms (like multigrid ones) consists in successively solving the problems: *for a given $f_i \in M_i$, find $v_i \in M_i$ such that*

$$L_i v_i = f_i, \quad i = 0, 1, \ldots, l. \tag{5}$$

Now let us formulate the cascadic conjugate-gradient algorithm (CCG-algorithm) in more detail.

CCG-algorithm.

    { 1. Set $u_0 = L_0^{-1} f_0$.

      2. For $i = 1, 2, \ldots, l$ do:

        { 2.1. Set $w_i = I_{i-1} u_{i-1}$;

        2.2. Do m_i iterations of the conjugate-gradient method:

$$y_0 = w_i;$$
$$p_0 = r_0 = f_i - L_i y_0;$$
$$\sigma_0 = (r_0, r_0)_i;$$
$$\text{for } k = 1, 2, \ldots, m_i \text{ do:}$$
$$\{ \quad \alpha_{k-1} = (p_{k-1}, L_i p_{k-1})/\sigma_{k-1};$$
$$y_k = y_{k-1} + \frac{1}{\alpha_{k-1}} p_{k-1};$$
$$r_k = r_{k-1} - \frac{1}{\alpha_{k-1}} L_i p_{k-1};$$
$$\sigma_k = (r_k, r_k)_i; \qquad \beta_k = \sigma_k/\sigma_{k-1};$$
$$p_k = r_k + \beta_k p_{k-1};$$
$$\} \quad \text{the end of the iteration;}$$

      2.3. Set $u_i = y_{m_i}$;

} the end of the level i;

} the end of the algorithm.

We shall study the convergence of the CCG-algorithm in the full symmetric case under the following assumptions:

$$\text{the operators } L_i \text{ are self-adjoint and positive-definite} \tag{6}$$

and

$$L_{i-1} = I_{i-1}^* L_i I_{i-1} \tag{7}$$

$\forall i = 1, 2, \ldots, l$ in the sense of inner product $(\cdot , \cdot)_i$, i.e., $\forall u, v \in M_i, w \in M_{i-1}$ we assume that

$$(L_i u, v)_i = (u, L_i v)_i;$$

$$(L_i u, u)_i \geq \alpha_i (u, u)_i, \quad \alpha_i > 0;$$

$$(I_{i-1}^* v, w)_{i-1} = (v, I_{i-1} w)_i.$$

Let us introduce the scale of norms for elements $u \in M_i$ by

$$|||u|||_i^{(\alpha)} = (L_i^\alpha u, u)_i$$

with $\alpha \in (-\infty, \infty)$. For ease of presentation, we mark out the norms for vectors

$$|||u|||_i = |||u|||_i^{(1)}, \quad ||u||_i = |||u|||_i^{(0)},$$

and for operators $B : M_i \to M_i$

$$|||B|||_i = \sup_{u \in M_i \setminus \{0\}} \frac{|||Bu|||_i}{|||u|||_i}. \tag{8}$$

Let us fix for a moment an integer $i \in [1, l]$ and denote by $\delta_0 = v_i - w_i$ the error of an initial guess w_i for the exact solution v_i of problem (5) and by $\delta_1 = v_i - u_i$ the error of the final approximation u_i. In such a way, we have defined the operator $B_i : \delta_0 \to \delta_1$ on M_i, called the operator of error reduction at the level i. Due to [2], this linear operator can be represented as a polynomial in L_i :

$$B_i = p_i(L_i) = I + \sum_{k=1}^{m_i} a_k L_i^k \tag{9}$$

with coefficients which depend on the parameters $\sigma_0, \ldots, \sigma_{m_i}, \alpha_0, \ldots, \alpha_{m_i-1}$. Here and further we denote by I the identity operator in a corresponding space.

Lemma 1 [2] *Under a fixed initial guess w_i (and respectively under a fixed initial error δ_0) the conjugate-gradient method minimizes the error norm $|||\delta_1|||_i$ of the final approximation u_i among all polynomials of the form (9) with arbitrary coefficients a_k.*

3 The main result

Let λ_i^* be the maximal eigenvalue of the operator L_i in the space M_i :

$$L_i\phi_j = \lambda_j\phi_j, \quad j = 1,\ldots,n_i = \dim M_i. \tag{10}$$

Let us introduce the polynomial in $x \in \mathbf{R}$

$$q_i(x) = \prod_{k=1}^{m_i} (1 - \mu_k x) \tag{11}$$

with parameters

$$\mu_k = \lambda_i^{*-1} \cos^{-2} \frac{\pi(2k-1)}{2(2m+1)}. \tag{12}$$

As it has been shown in [3], this polynomial has the least deviation from zero

$$\max_{x\in[0,\lambda_i^*]} |\sqrt{x}q_i(x)| = \frac{\sqrt{\lambda_i^*}}{2m_i + 1} \tag{13}$$

among all polynomials of the form $p_i(x)$ from (9). Another useful property of this polynomial consists in the inequality

$$|q_i(x)| \leq 1 \quad \text{on } [0, \lambda^*]. \tag{14}$$

Now let us introduce the matrix polynomial

$$Q_i = q_i(L_i) = \prod_{k=1}^{m_i} (I - \mu_k L_i). \tag{15}$$

Lemma 2 *Let (6) hold. Then $\forall \alpha \in (0,1]$*

$$|||Q_i w|||_i \leq \frac{(\lambda_i^*)^{\alpha/2}}{(2m_i + 1)^\alpha} |||w|||_i^{(1-\alpha)} \qquad \forall w \in M_i. \tag{16}$$

Now we formulate the convergence criterion: *there exist constants $c^* > 0$ and $\alpha \in (0,1]$ such that $\forall i = 1,\ldots,l$*

$$|||v_i - I_{i-1}v_{i-1}|||_i^{(1-\alpha)} \leq c^*(\lambda_i^*)^{-\alpha/2}|||v_i - I_{i-1}v_{i-1}|||_i. \tag{17}$$

So we can prove the main estimate by analogy with Theorem 4.1 from [6].
Theorem. *Let (6), (7), (17) hold. Then*

$$|||v_i - u_i|||_i \leq c^* \sum_{j=1}^{i} \frac{1}{(2m_j + 1)^\alpha}|||v_j - I_{j-1}v_{j-1}|||_j. \tag{18}$$

4 The verification of algebraic properties for the finite-element formulation

Let us consider the Dirichlet problem in a polygon $\Omega \subset \mathbf{R}^2$ with a boundary Γ:

$$-\sum_{i,j=1}^{2} \partial_i(a_{ij}\partial_j u) + au = f \qquad \text{in} \quad \Omega, \tag{19}$$

$$u = 0 \qquad \text{on} \qquad \Gamma, \tag{20}$$

where the coefficients and the right-hand side of (19) satisfy the conditions

$$\partial_k a_{ij} \in L_q(\Omega), \quad q > 2, \quad i,j,k = 1,2; \qquad a_{12} = a_{21} \text{ on } \bar{\Omega};$$

$$\mu \sum_{i=1}^{2} \xi_i^2 \leq \sum_{i,j=1}^{2} a_{ij}\xi_i\xi_j \leq \nu \sum_{i=1}^{2} \xi_i^2 \quad \text{on} \quad \bar{\Omega} \quad \forall\, \xi_i \in \mathbf{R}, \; \nu \geq \mu > 0; \tag{21}$$

$$a, f \in L_2(\Omega); \qquad a \geq 0 \quad \text{on} \quad \bar{\Omega}.$$

These conditions quarantee that the problem (19), (20) has a unique solution in the class $H^{1+\alpha}(\Omega)$, moreover, the estimate

$$\|u\|_{1+\alpha,\Omega} \leq c_1 \|f\|_{\alpha-1,\Omega} \tag{22}$$

holds with some $\alpha \in (0,1]$. We use the common notation of the Sobolev spaces $H^s(\Omega)$ and norms $\|\cdot\|_{s,\Omega}$ with $s \in (-\infty, \infty)$. If polygon Ω is convex, all interior angles are less than π and we have the standard regular case with $\alpha = 1$ which is studied in [6].

Let us treat a case with a non-convex polygon. To ease of presentation, we take a poligon Ω whose interior angles are less than π except one angle at the origin $(0, 0)$ with a value greater than π. Then α in (22) is less than 1 [3]: $\alpha \in (0,1)$. The problem (19), (20) may be reduced to the weak formulation: *find* $u \in H_0^1(\Omega)$ *such that*

$$\mathcal{L}(u,v) = (f,v)_\Omega \qquad \forall v \in H_0^1(\Omega), \tag{23}$$

where the bilinear forms are given by

$$\mathcal{L}(u,v) = \int_\Omega \Big(\sum_{i,j=1}^{2} a_{ij}\partial_j u \partial_i v + auv\Big)dx, \quad (f,v)_\Omega = \int_\Omega uv dx.$$

Due to (21), this problem has a unique solution too [4]. To construct the Bubnov-Galerkin scheme, we first divide the initial polygon Ω into a small amount of closed triangles so that this triangulation is admissible, i.e., each pair

190

of triangles has either no common points, or a common vertex, or a common side. Denote the maximal length of the sides of all triangles by h_0. Put $N_i = 2^i, h_i = h_0/N_i$. Then we consider two different constructions of triangulations.

4.1   The piecewise uniform triangulation

The first one gives a piecewise uniform triangulation. For all $i = 1, ..., l$, let us divide every initial triangle into N_i^2 equal triangles. Denote the set of all vertices of the obtained admissible triangulation by $\bar{\Omega}_i$ and introduce the set $\Omega_i = \bar{\Omega}_i \cap \Omega$ and the number n_i of nodes of Ω_i. For every node $y \in \Omega_i$, let us introduce the basis function $\varphi_y^i \in H_0^1(\Omega)$ which is linear on each triangle of triangulation $\bar{\Omega}_i$, equals 1 at the node y and equals 0 at any other node $z \in \bar{\Omega}_i$. Denote the linear span of these functions

$$\mathcal{H}^i = \quad \text{span} \quad \{\varphi_y^i\}, \quad y \in \Omega_i.$$

Restricting (23) into the subspace $\mathcal{H}^i \in H_0^1(\Omega)$, we get the discrete problem: find $\tilde{v}_i \in \mathcal{H}^i$ such that

$$\mathcal{L}(\tilde{v}_i, v) = (f, v)_\Omega \quad \forall v \in \mathcal{H}^i. \tag{24}$$

Let M_i be the n_i-dimensional space of vectors w with components $w(x), x \in \Omega_i$. Then formulation (24) is equivalent to the linear system of algebraic equations (5) with the symmetric matrix L_i and the right-hand side f_i with the elements $L_i(x, y) = \mathcal{L}(\varphi_x^i, \varphi_y^i)$, $f_i(x) = (f, \varphi_x^i)_\Omega, x, y \in \Omega_i$. Let us define the usual isomorphism between vectors $v \in M_i$ and functions $\tilde{v} \in \mathcal{H}^i$ which are their prolongations, i.e.,

$$\tilde{v}(x) = \sum_{y \in \Omega_i} v(y)\varphi_y^i(x), \quad x \in \bar{\Omega}, \tag{25}$$

and vica versa

$$v(y) = \tilde{v}(y), \quad y \in \Omega_i. \tag{26}$$

Now we introduce the energy norm for functions:

$$|||v|||_\Omega = \mathcal{L}(v, v)^{1/2}, \quad v \in H_0^1(\Omega), \tag{27}$$

and specify the inner product and the norm for vectors:

$$(v, w)_i = \sum_{x \in \Omega_i} v(x)w(x)h_i^2, \quad \|v\|_i = (\sum_{x \in \Omega_i} v^2(x)h_i^2)^{1/2}, \quad v, w \in M_i. \tag{28}$$

Due to (25),(26), we have

$$|||v|||_i = h_i |||\tilde{v}|||_\Omega \tag{29}$$

Due to (25),(26), we have

$$|||v|||_i = h_i|||\tilde{v}|||_\Omega \tag{29}$$

for an isomorphic pair $v \in M_i, \tilde{v} \in \mathcal{H}^i$. L^2-norm $\|\tilde{v}\|_{0,\Omega}$ is not equal to $\|v\|_i$ but is equivalent to it [3]:

$$c_2\|v\|_i \leq \|\tilde{v}\|_{0,\Omega} \leq c_3\|v\|_i. \tag{30}$$

From (29) and (30), the inequality

$$c_2'|||v|||_i^{(s)} \leq h_i^s\|\tilde{v}\|_{s,\Omega} \leq c_3'|||v|||_i^{(s)}, \quad s \in [0,1], \tag{31}$$

follows, which is proved in [7] in other terms.

Now let us introduce the interpolation operator $I_i : M_i \to M_{i+1}$. Let $v \in M_i$. Since its prolongation $\tilde{v}$ belongs to $\mathcal{H}^{i+1}$, the isomorphism gives us a vector $w \in M_{i+1}$ for $\tilde{v}$. In such a way, we uniquely defined $I_i : v \to w$.
The covergence of the Bubnov-Galerkin solution to the exact one was studied in a number of paper (e.g., [5]).

Lemma 3 *Under conditions (21) on the piecewise uniform triangulation, the solution of (24) exists, is unique and obeys the estimate*

$$|||u - \tilde{v}_i|||_\Omega \leq c_4 h_i^\alpha \|f\|_{\alpha-1,\Omega}. \tag{32}$$

Note, that we have the usual estimate

$$0 < \lambda_i^* \leq c_6 \quad \forall \quad i = 1,\dots,l \tag{33}$$

for the largest eigenvalue λ_i^* of L_i (see e.g., [3]). To check the convergence criterion (17), we use the Nitsche trick and inequality (31).

Lemma 4 *Under the conditions (21) and (22) on the piecewise uniform triangulation, the criterion (17) is valid with α from (22).*

So, all assumptions of Theorem are valid and we have the estimate (18). A more convinient form follows from the inequality

$$|||\tilde{v}_j - \tilde{v}_{j-1}|||_\Omega \leq |||u - \tilde{v}_{j-1}|||_\Omega \tag{34}$$

which is proved in [6]. It permits to transform the estimate (18):

$$|||v_i - u_i|||_i \leq c^* \sum_{j=1}^i \frac{1}{(2m_j + 1)^\alpha}|||u - \tilde{v}_{j-1}|||_\Omega \tag{35}$$

or due to (32)

$$|||v_i - u_i|||_i \leq c^* c_4 \sum_{j=1}^i \frac{h_{j-1}^\alpha}{(2m_j + 1)^\alpha}\|f\|_{\alpha-1,\Omega}. \tag{36}$$

192

4.2 The special refined triangulation

Now we discuss the second construction with a special refinement of triangulation near singular points. For example, one can implement a technique described in [3]. We begin with the initial admissible triangulation $\bar{\Omega}_0$ again. Every initial triangle is further divided into 4 finer ones. To this end, we introduce a refinement index $\rho \geq 1$. If a triangle of i-th triangulation does not touch to the origin $(0, 0)$, it will be broken into 4 equal triangles of the $(i+1)$-th triangulation by connecting midpoints of its sides. Consider now a triangle $\triangle ABC$ of the i-th triangulation with vertex $A = (0,0)$. Denote by b the length of the height from A to the opposite side of an initial triangle. Then we draw a straight segment $B'C'$ parallel to BC at a distance $bn_i^{-\rho}$ from A with ends B', C' on the sides AB, AC correspondingly. Denote the midpoint of BC by A'. Joining the points A', B', C' by straight segments gives 4 finer triangles of the $(i+1)$-th triangulation. By breaking in such way consistently for $i = 1, \ldots, l$, we get admissible triangulations with the vertices set $\bar{\Omega}_i$. We keep definitions of $\Omega_i, n_i, \varphi_y^i, \mathcal{H}^i$ and (27), (29), (30). But for an optimal rate of convergence, we change the definitions (28). Each node $x \in \Omega_i$ corresponds to the value d_x^i, a total area of triangles of i-th triangulation with the vertex x. Let us construct the diagonal matrix $D_i = diag\{d_x^i\}, x \in \Omega_i$. Instead of (28), we introduce the inner product and the norm

$$(v,w)_i = \sum_{x \in \Omega_i} v(x)w(x)d_x^i, \quad \|v\|_i = (\sum_{x \in \Omega_i} v^2(x)d_x^i)^{1/2}, \quad v, w \in M_i. \qquad (37)$$

If we formulate the systems (5) in a standard way, their matrices are not self-adjoint in the inner product (37). Therefore we shall formulate these systems with

$$L_i = h_i^2 D_i^{-1} \tilde{L}_i, \quad f_i = h_i^2 D_i^{-1} \tilde{f}_i, \qquad (38)$$

where $\tilde{L}_i, \tilde{f}_i$ are the standard matrix and the right-hand side with the elements $\tilde{L}_i(x,y) = \mathcal{L}(\varphi_x^i, \varphi_y^i)$, $\tilde{f}_i(x) = (f, \varphi_x^i)_\Omega$, $x, y \in \Omega_i$. For these matrices, we have (29), (30) again. But the estimate (33) becomes worse [3]:

$$0 < \lambda_i^* \leq c_5 h_i^{2-2\rho} \quad \forall\, i = 1, \ldots, l. \qquad (39)$$

Now we reformulate the previous lemmata for this type of triangulation.

Lemma 5 *[3]. Under conditions (21) on the refined triangulation for any $\rho \geq 1$, the solution of (24) exists and is unique. There exists $\rho_0 > 1$ such that $\forall\, \rho \geq \rho_0$*

$$|||u - \tilde{v}_i|||_\Omega \leq c_5 h_i \|f\|_{0,\Omega}. \qquad (40)$$

Lemma 6 *Under conditions (21) on the refined triangulation with $\rho = \rho_0$, the criterion (17) is valid with $\alpha = 1/\rho_0$.*

Since all assumptions of Theorem are valid, we have the estimates (18) and (35). Instead of (36), we get

$$|||v_i - u_i|||_i \leq c^* c_5 \sum_{j=1}^{i} \frac{h_{j-1}}{(2m_j + 1)^\alpha} \|f\|_{0,\Omega}. \tag{41}$$

5 An evaluation of the amount of iterations

From the formulae (36) and (41), we see the following. To get an acceptable accuracy of u_i, we have to increase the amount m_j of iterations in CCG-algorithm from top to bottom. The more h_{j-1}, the more m_j. For example, one can choose an exponential dependence with some constants $\bar{m} = m_l, \sigma$:

$$\sigma(h_i/h_l)^\beta(2\bar{m} + 1) \leq 2m_i + 1 \leq (h_i/h_l)^\beta(2\bar{m} + 1), \tag{42}$$
$$i = 0, 1, ..., l - 1; \quad \sigma > 1, \quad \text{e.g.,} \, \sigma = 1.2.$$

If $\beta > 1$, we get

$$|||v_l - u_l|||_l \leq c_\beta \frac{h_l^\gamma}{(2\bar{m} + 1)^\alpha} \|f\|_{\gamma-1,\Omega} \tag{43}$$

from (36) and (41), where $\gamma = \alpha < 1$ for the piecewise uniform triangulation and $\gamma = 1$ for the special refined one. We see that the acceptable accuracy (which is comparable with the discretization error) can be achieved by a finite number $\bar{m}$.

Let us evaluate the amount N_{CCG} of arithmetic operations in full CCG-algorithm. If we take $\beta < 2$,

$$N_{CCG} \leq c_\beta'(\bar{m} + c_9)n_l. \tag{44}$$

I.e., full CCG-algorithm has a finite amount of arithmetic operations per one unknow of the system at the highest level.

The question arises: What value of β is optimal? Applying the method of Lagrange multipliers, we obtain [6] the following dependence:

$$(2m_i + 1) = (2\bar{m} + 1)(n_l h_{i-1}^\gamma / n_i h_{l-1}^\gamma)^{1/(\alpha+1)}.$$

References

[1] Deuflhard, P.: Cascadic Conjugate Gradient Methods for Elliptic Partial Differential Equations I. Algorithm and Numerical Results.– Technical Report SC 93-23. Konrad-Zuse-Zentrum Berlin (ZIB), 1993.

[2] Samarskiĭ, A. A., Nikolaev, E. S.: Numerical Methods for Grid Equations. V. II, Iterative Methods.– Berlin: Birkhauser 1989.

[3] Shaĭdurov, V. V.: Multigrid Methods of Finite-Elements.–Amsterdam: Kluver 1994; the Russian version: Moscow: Nauka 1989.

[4] Ladyzhenskaya, O. A., Uraltseva N. N.: Linear and Quasilinear Equations of Elliptic Type. – Moscow: Nauka 1973.

[5] Ciarlet, P.: The Finite Element Method for Elliptic Problems.– Amsterdam: North-Holland 1978.

[6] Shaĭdurov, V. V.: Some Estimates of the Rate of Convergence for the Cascadic Conjugate-Gradient Method.– Technical Report. Otto-von-Guericke-Universität Magdeburg 1994.

[7] Bank, R. E., Dupont, T. F.: An Optimal Order Process for Solving Elliptic Finite Element Equations.– Math. Comp., 1981, v.36, p.35–51.

Vladimir Shaĭdurov*
Computing Center of
Russian Academy of Sciences
Akademgorodok Krasnoyarsk
Russia 660036
e-mail: shidurov@intr.sibch.glas.apc.org

---

*Research was supported by Institut für Angewandte Analysis und Stochastik im Forschungsverbund Berlin e.V. and by Institut für Analysis und Numerik, Otto-von-Guericke-Universität Magdeburg.

Selfadaptive FE–approach of shell buckling and postcritical solution branches including dimensional recovery in disturbed subdomains

Erwin Stein, Bengt Seifert, Stephan Ohnimus

1   Introduction

Moderate and finite rotation theories [7] [4] [6] are available including the finite element analysis, especially the treatment of shear locking problems. The five parameter shell model used in this paper is based on geometrically non–linear Reissner–Mindlin kinematics where the finite element formulation may be derived using the Biot- or 2. Piola- Kirchhoff stress resultants, alternatively. The description of finite rotations of the shell normal is expressed in terms of a scew–symmetric tensor.

The reliable computation of geometrical non–linear load displacement paths needs a suitable error controlled refinement strategy, especially for the treatment of bifurcation problems. Those adaptive methods for detecting local and global buckling need consistent and stable modelling in conjunction with numerical approximations. Additionally dimensional adaptivity is taken into account to take care for 3D- disturbances of the 2D-model, mainly at supports.

Three numerical examples are presented in order to show the effectivity of the a posteriori error indicators, to give a deeper insight into the investigated shell problems and to illustrate the proposed solutions strategies.

2 A thin shell element for small elastic strains and finite rotations

In this section we discuss a five parameter finite element formulation applicable to structural problems with finite rotations. The St.Venant–Kirchhoff hyperelastic material law restricts its application to small strain problems. A displacement approach is used which includes shear deformation, and therefore belongs to non–linear Reissner-Mindlin type theories. The kinematics for the finite element formulation are derived from the description of the movement of an orthonormal cartesian frame during deformation, identical with the initial normal vector and the tangents in the undeformed state. Since shear deformation is included the rotated director is not normal to the deformed middle surface, and therefore the rotated base vectors do not remain tangents with respect to the deformed surface.

2.1 Kinematics of the shell and variational formulation

The usual kinematical setting describes the position of an arbitrary point in the shell space where $\mathbf{x}$ denotes its position vector in the current configuration, see Fig.1. Here

$$\mathbf{x} = \mathbf{x}_0 + s_3 \mathbf{a}_3 , \quad \mathbf{x}_0 = \mathbf{X}_0 + \mathbf{u} = (X_{0i} + u_i)\mathbf{e}_i , \quad -\frac{h}{2} \le s_3 \le \frac{h}{2} \quad . \tag{1}$$

describes the translation of the shell middle surface from its initial into the current configuration whereas s_3 is the coordinate in the thickness direction and $\mathbf{a}_3$ the shell director in the current configuration.

Furthermore we introduce an orthonormal coordinate system $\mathbf{t}_i$ in the reference configuration where the vectors $\mathbf{t}_1$ and $\mathbf{t}_2$ are tangents to the shell middle surface, and s_1 and s_2 are the associated coordinates. An additional orthonormal basis is constructed in the current configuration of the shell, and the following relations between the two different bases hold.

$$\mathbf{t}_i(s_\alpha) = \mathbf{R}_0(s_\alpha)\,\mathbf{e}_i , \quad \mathbf{a}_i(s_\alpha) = \mathbf{R}(s_\alpha)\,\mathbf{t}_i , \quad \alpha = 1,2 . \tag{2}$$

Since the two matrices $\mathbf{R}_0$ and $\mathbf{R}$ describe an orthogonal transformation the Euclidean norm of the director vector $\mathbf{a}_3$ remains unchanged during deformation.

The following expressions for the derivatives of the base vectors are needed for subsequent development

$$\frac{\partial t_i(s_\alpha)}{\partial s_\alpha} = \Omega_{0\alpha} t_i = \omega_{0\alpha} \times t_i \ , \qquad \Omega_{0\alpha} = \frac{\partial R_0(s_\alpha)}{\partial s_\alpha} R_0^T \qquad (3)$$

$$\frac{\partial a_i(s_\alpha)}{\partial s_\alpha} = (\Omega_\alpha + R\Omega_{0\alpha}R^T)a_i = (\omega_\alpha + R\omega_{0\alpha}) \times a_i \ , \qquad \Omega_\alpha = \frac{\partial R(s_\alpha)}{\partial s_\alpha} R^T \ .$$

$$(4)$$

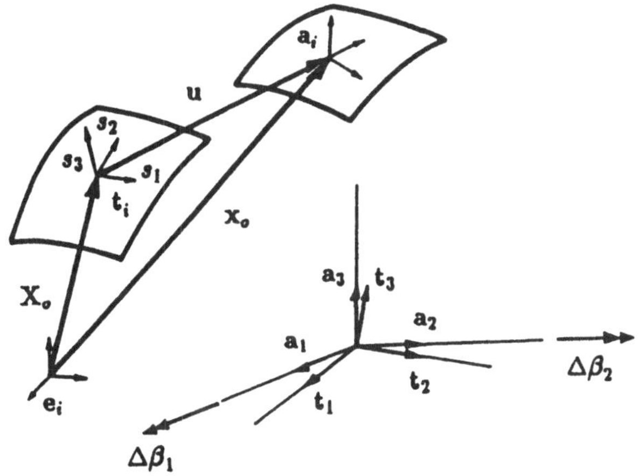

Fig. 1 Reference and deformed middle surface of the shell

The scew-symmetric tensors $\Omega_{0\alpha}$ and Ω_α are related to associated axial vectors $\omega_{0\alpha}$ and ω_α which are necessary for Finite Element implementation. Stress resultants and stress couple resultants are defined by integration of the stress vectors t_α across the shell thickness

$$n_\alpha = \int_{-\frac{h}{2}}^{\frac{h}{2}} t_\alpha ds_3 \ , \qquad m_\alpha = \int_{-\frac{h}{2}}^{\frac{h}{2}} (x - x_0) t_\alpha ds_3 \ , \qquad (5)$$

where the stress vectors t_α are related to the 1. Piola Kirchhoff tensor $P = t_i \otimes e_i$. Exploiting standard arguments of variational calculus [4] the equilibrium equations are multiplied with statically admissible variations and yield the associated virtual work in terms of Biot stress resultants $N_\alpha = R_0^T R^T m_\alpha$, $M_\alpha = R_0^T R^T n_\alpha$ and related work conjugated strain measures. Another transformation yields symmetrized 2. Piola Kirchhoff stress resultants $\tilde{N}_{\alpha\beta}$, the stress couple resultants $\tilde{M}_{\alpha\beta}$ and transverse shear forces $\tilde{Q}_\alpha$ as well as the

198

work conjugated Green Lagrangian membrane strains $\varepsilon_{\alpha\beta}$, shear strains γ_α and changes of curvatures $\kappa_{\alpha\beta}$

$$\varepsilon_{\alpha\beta} \;=\; \frac{1}{2}\,(\,\mathbf{x}_{o,\alpha}\cdot\mathbf{x}_{o,\beta} \;-\; \mathbf{X}_{o,\alpha}\cdot\mathbf{X}_{o,\beta}\,) \tag{6}$$

$$\gamma_\alpha \;=\; \mathbf{x}_{o,\alpha}\cdot\mathbf{a}_3 \;-\; \mathbf{X}_{o,\alpha}\cdot\mathbf{t}_3 \tag{7}$$

$$\kappa_{\alpha\beta} \;=\; \mathbf{x}_{o,\alpha}\cdot\mathbf{a}_{3,\beta} \;-\; \mathbf{X}_{o,\alpha}\cdot\mathbf{t}_{3,\beta}\;. \tag{8}$$

The virtual work of the internal forces then reads

$$\delta W = \int_\Omega (\,\tilde{N}_{\alpha\beta}\,\delta\varepsilon_{\alpha\beta} \;+\; \tilde{M}_{\alpha\beta}\,\delta\kappa_{\alpha\beta} \;+\; \tilde{Q}_\alpha\,\delta\gamma_\alpha\,)\,d\Omega \tag{9}$$

which is used for the finite element approximation.

2.2 Geometry interpolation and finite element discretization

In this section the finite element formulation for the theory described above is summarized which involves only axial vectors and the rotation tensor. The chosen unknowns for the rotations are the components of the axial pseudo vector $\Delta\boldsymbol{\psi}^i$, where the superscript i denotes a single step of the Newton iteration. The associated skew-symmetric matrix $\Delta\boldsymbol{\Omega}^i$ results from a linearization of the incremental rotation tensor, i.e. a Taylor expansion of the tensor $\Delta\mathbf{R}^i$. This parametrization is supposed to be singularity free. The update of the whole rotation tensor $\mathbf{R}^i$ within each step of the Newton iteration is computed exactly with the incremental tensor $\Delta\mathbf{R}^i$ [6].
An isoparametric C^0 continuous approach is used, and five nodal values (subscript k) are interpolated with bilinear interpolation funtions (N_k).
The incremental nodal displacement vector is defined as

$$\Delta\mathbf{V}_k^i = [\Delta v_{1k}^i, \Delta v_{2k}^i, \Delta v_{3k}^i, \Delta\beta_{1k}^i, \Delta\beta_{2k}^i]$$

$$\Delta\mathbf{x}^i = \sum_k N_k \Delta\mathbf{x}_k^i\;, \qquad \Delta\mathbf{u}^i = \sum_k N_k \Delta\mathbf{v}_k^i\;, \qquad \Delta\boldsymbol{\psi}^i = \sum_k N_k \Delta\boldsymbol{\psi}_k^i\;. \tag{10}$$

Two independent local rotation angles $\Delta\boldsymbol{\beta}^i$ are transformed into the three global rotation angles $\Delta\boldsymbol{\psi}^i$ with the matrix $\boldsymbol{\Lambda}$

$$\Delta\boldsymbol{\psi}^i = \boldsymbol{\Lambda}\,\Delta\boldsymbol{\beta}^i\;, \qquad \boldsymbol{\Lambda} = [\mathbf{a}_1|\mathbf{a}_2]\;. \tag{11}$$

At nodal points the update of the displacement field is obtained additively

$$\mathbf{u}^{i+1} = \Delta\mathbf{u}^i + \mathbf{u}^i \ , \tag{12}$$

whereas the rotation field needs a multiplicative update starting from the initial position of the cartesian frame

$$\mathbf{R}^0 = [\mathbf{t}_1|\mathbf{t}_2|\mathbf{t}_3] \ , \quad \mathbf{R}^{i+1} = \Delta\mathbf{R}^i \ \mathbf{R}^i \ , \quad \mathbf{R}^i = [\mathbf{a}_1|\mathbf{a}_2|\mathbf{a}_3]^i \ . \tag{13}$$

The following expression for the Taylor expansion of $\Delta\mathbf{R}^i$ (Rodriguez formula) can befound in [6], e.g.

$$\Delta\mathbf{R}^i = 1 + \frac{sin\|\Delta\boldsymbol{\psi}^i\|}{\|\Delta\boldsymbol{\psi}^i\|}\Delta\boldsymbol{\Omega}^i + \frac{1}{2}\frac{sin^2\|\frac{\Delta\boldsymbol{\psi}^i}{2}\|}{\|\frac{\Delta\boldsymbol{\psi}^i}{2}\|^2}(\Delta\boldsymbol{\Omega}^i)^2 \ . \tag{14}$$

The construction of the discrete variational form is performed with the expressions given for $\Delta\boldsymbol{\psi}^i$, $\Delta\boldsymbol{\Omega}^i$ and $\mathbf{R}^i$. The following standard linearization process (e.g. [6]) of the discrete static weak form yields the tangent operator $\mathbf{K}^i$ and the residuum $\mathbf{G}^i$.

3 Implementation of Rheinboldt's error indicator

A further analysis of Rheinboldt's linearized indicator [5], see also Verfürth [10], yields conditions under which linearized error indicators are reliable, see [9]. Following non–linear load deflection curves a non–linear smooth mapping $\mathbf{G} : \mathbf{X} \mapsto \mathbf{X}^*$, $\mathbf{G}(\mathbf{u}) = \lambda \cdot \mathbf{p}$, maps displacements $\mathbf{u}$ from a real vector space $\mathbf{X}$ onto its dual space of forces $\mathbf{X}^*$. The real vector space is a Banach space and equipped with a suitable norm, e.g. $\|\cdot\|$ the energy norm. Linearization at an equilibrium point $(\mathbf{u}_0, \lambda_0)$, which satisfies the equilibrium equation above exactly, yields $\mathbf{K}_T(\mathbf{u}_o)\Delta\mathbf{u} = -\Delta\mathbf{G}(\mathbf{u}_o)$. $\mathbf{K}_T(\mathbf{u}_o) := D\mathbf{G}(\mathbf{u}_0)$ denotes the Frechet–derivative of $\mathbf{G}$. Furthermore two assumptions, namely $\|D\mathbf{G}(\mathbf{u})^{-1}\| \leq M_1$ and $\|D^2\mathbf{G}(\mathbf{u})\| \leq M_2$, with any allowable $\mathbf{u} \in \mathbf{X}$ which satisfies $\|\mathbf{u} - \mathbf{u}_0\| \leq \delta_2$, are needed. According to [9] the error of the real solution $\tilde{\mathbf{u}}_0 - \mathbf{u}_0$ can be evaluated by the exact solution of the linearized problem $\mathbf{w}_0$ as

$$\|\tilde{\mathbf{u}}_0 - \mathbf{u}_0\| \cdot (1 + c) = \|\tilde{\mathbf{u}}_0 - \mathbf{w}_0\| \ , \quad |c| \leq \frac{1}{2}M_1 M_2 \delta_2 \tag{15}$$

200

for all approximations $\tilde{\mathbf{u}}_0$. Thus the error $\|\tilde{\mathbf{u}}_0 - \mathbf{u}_0\|$ is nearly equal to the difference between the approximation and the solution of the linearized equation $\|\tilde{\mathbf{u}}_0 - \mathbf{w}_0\|$. It suffices to estimate the latter one, $\|\tilde{\mathbf{u}}_0 - \mathbf{w}_0\|$, with respect to the linearized problem. Asymptotic behaviour can only be achieved when the constant c is small, which is not a priori clear in practical situations. Therefore this must also be controlled within the iterative mesh refinement algorithm. For instance the constant c is far from beeing bounded for locking dominated problems.

In the case of bifurcation problems M_1 is large in a neighbourhood of a stability point $(\mathbf{u}^*, \lambda^*)$. Hence it is not a–priori clear that c is small. Thus, mesh adaptivity cannot improve the accuracy of a critical load within an ϵ–surrounding. Corresonding problems arise for the eigenvectors and lead to difficulties in case of clustering eigenvalues.

In order to obtain incremental stress states, the Frechet directional derivative yields

$$\Delta \boldsymbol{\sigma}_h = \frac{\partial \boldsymbol{\sigma}}{\partial \mathbf{v}} \Delta \mathbf{v}_{trial} = \mathbf{C}\,\mathbf{B}(\mathbf{v})\,\Delta \mathbf{v}_{trial} \tag{16}$$

where different choices of $\mathbf{v}_{trial}$ are possible. For instance within a certain neighbourhood of bifurcation points measured with the norm $\| \mathbf{u}^* - \mathbf{u}^\circ \|$, the normalized eigenvector $\boldsymbol{\Phi}$ according to the lowest eigenvalue of $\mathbf{K}_T$ is used for applying the error estimator. The first eigenvalue may be very small but not zero. In this case we obtain the incremental displacement

$$\Delta \mathbf{v}_{trial} = \boldsymbol{\Phi}\left(\mathbf{K}_T\left(\mathbf{v}\left(\lambda_{crit}\right)\right)\right) \tag{17}$$

where single valued bifurcations are assumed. Due to the results of [9], adaptivity is only possible before and after bifurcations.

The residual Babuska – Miller error indicator [1] is adequate for the linearized elastic problem, i.e. checking $\Delta \boldsymbol{\sigma}_h$. For bilinear shape functions only the stress jumps at element boundaries are relevant.

For a shell element e we get the a–posteriori indicator with respect to its neighbours

$$\eta^2_{e\,BM} = \eta^2_{e\,MEM} + \eta^2_{e\,BEN} + \eta^2_{e\,SHEAR} \tag{18}$$

$$\eta^2_{e\,MEM} = \frac{t_e}{K^2_{e\,MEM}} \int_{\partial\Omega_e} \mathbf{J}\left(\Delta \mathbf{n}\right)^T \mathbf{J}\left(\Delta \mathbf{n}\right) ds \tag{19}$$

$$\eta^2_{e\,BEN} = \frac{t^3_e}{K^2_{e\,BEN}} \int_{\partial\Omega_e} \mathbf{J}\left(\Delta \mathbf{m}\right)^T \mathbf{J}\left(\Delta \mathbf{m}\right) ds \tag{20}$$

$$\eta^2_{e\,SHEAR} \;=\; \frac{t_e}{K^2_{e\,SHEAR}} \int_{\partial\Omega_e} \mathbf{J}\,(\Delta\mathbf{q})^T\,\mathbf{J}\,(\Delta\mathbf{q})\,ds \qquad (21)$$

with

$$K_{BEN} = \frac{Et_e^3}{12(1-\nu^2)}\,, \quad K_{MEM} = \frac{Et_e}{1-\nu^2}\,, \quad K_{SHEAR} = \kappa\,\frac{Et_e}{2(1+\nu)} \qquad (22)$$

4 Aspects of dimensional adaptivity in disturbed 2D–subdomains

4.1 Introduction and concepts

Modeling of thin–walled beams, plates and shells as 1D–and 2D–BVP's due to kinematical hypotheses, e.g. the normal hypothesis, is valid in undisturbed domains. Disturbances near supports and free edges, in the vicinity of concentrated loads and at thickness jumps cannot be described by 1D– and 2D–BVP's. In these disturbed subdomains dimensional adaptivity has to be performed with respect to the 3D–theory. Dimensional adaptivity (d–adaptivity) coupled with an h–adaptivity or p-adaptivity (depending from the regularity of the solution) becomes necessary in oder to guarantee a reliable overall solution. Some results from dimensional adaptivity are published in [7] [8], including expansion and reduction methods. In this paper we discuss the dimensional adaptivity with anisotropic– refinements, using scaled anisotropic residual error indicators in the energy norm.

4.2 Anisotropic residual error indicator with respect to skew convected coordinates

The results from the error analysis (see *Johnson* [3] and *Babuška* [2]) end up in the following local element error indicator

$$\eta^2_e = \|u - u_h\|^2_{E(\Omega_e)} = \|\frac{1}{2}c_J h^{1/2}\boldsymbol{J}\|^2_{L_2(\partial\Omega_e)} + \|c_R h\boldsymbol{R}\|^2_{L_2(\Omega_e)}. \qquad (23)$$

The residual $\boldsymbol{R} = \boldsymbol{f} + \mathrm{div}\boldsymbol{\sigma}(\boldsymbol{u}_h)$, where $\boldsymbol{f}$ are the volume forces, and $\boldsymbol{J}$ is the jump of the approximated stress vector, $+$ denotes the section with positive coordinate increments $\boldsymbol{J} = \boldsymbol{t}_h^+ - \boldsymbol{t}_h^-$ and $\boldsymbol{t} = \boldsymbol{\sigma}(\boldsymbol{u}_h)\cdot\boldsymbol{n}$. Note, that for implementation of the error indicator the first term of equ. (23) must be multiplied by 4 if the element boundary is equal to the system boundary, and the jump of the stress vector at the element boundary becomes $\boldsymbol{J} = \bar{\boldsymbol{t}} - \boldsymbol{\sigma}(\boldsymbol{u}_h)\cdot\boldsymbol{n}$. The method

202

for computing the interpolation constants c_J and c_R was given by *Johnson* [3]. The improvement of quantitative error anlysis is a topic of current research at our institute.

For determining the anisotropic error indicator we need for integration over the domain Ω_e an orthonormal basis $\boldsymbol{e}_r, \boldsymbol{e}_s, \boldsymbol{e}_t$ related to the convected skew base vectors

$$\tilde{\boldsymbol{g}}_r = \frac{d\boldsymbol{x}}{dr} \quad ; \quad \tilde{\boldsymbol{g}}_s = \frac{d\boldsymbol{x}}{ds} \quad ; \quad \tilde{\boldsymbol{g}}_t = \frac{d\boldsymbol{x}}{dt} \quad , \tag{24}$$

and an orthonormal basis $\bar{\boldsymbol{e}}_r, \bar{\boldsymbol{e}}_s, \bar{\boldsymbol{e}}_t$ for the integration across the element boundary $\partial\Omega_e$ with respect to the convected special skew basis at element surfaces

$$\tilde{\boldsymbol{g}}_r = \tilde{\boldsymbol{g}}_s \times \tilde{\boldsymbol{g}}_t \quad ; \quad \tilde{\boldsymbol{g}}_s = \frac{d\boldsymbol{x}}{ds} \quad ; \quad \tilde{\boldsymbol{g}}_t = \frac{d\boldsymbol{x}}{dt} \quad \text{for} \quad r = \{1, -1\} \quad , \tag{25}$$

$$\tilde{\boldsymbol{g}}_r = \frac{d\boldsymbol{x}}{dr} \quad ; \quad \tilde{\boldsymbol{g}}_s = \tilde{\boldsymbol{g}}_t \times \tilde{\boldsymbol{g}}_r \quad ; \quad \tilde{\boldsymbol{g}}_t = \frac{d\boldsymbol{x}}{dt} \quad \text{for} \quad s = \{1, -1\} \quad , \tag{26}$$

$$\tilde{\boldsymbol{g}}_r = \frac{d\boldsymbol{x}}{dr} \quad ; \quad \tilde{\boldsymbol{g}}_s = \frac{d\boldsymbol{x}}{ds} \quad ; \quad \tilde{\boldsymbol{g}}_t = \tilde{\boldsymbol{g}}_r \times \tilde{\boldsymbol{g}}_s \quad \text{for} \quad t = \{1, -1\} \quad . \tag{27}$$

Then the components of η_e are given by projections as [1]

$$\eta_{r_e}^2 = \|\tfrac{1}{2} c_{r_J} h_r^{1/2} \boldsymbol{J} \cdot \bar{\boldsymbol{e}}_r\|_{L_2(\partial\Omega_e)}^2 + \|c_{r_R} h_r \boldsymbol{R} \cdot \boldsymbol{e}_r\|_{L_2(\Omega_e)}^2 \tag{28}$$

$$\eta_{s_e}^2 = \|\tfrac{1}{2} c_{s_J} h_s^{1/2} \boldsymbol{J} \cdot \bar{\boldsymbol{e}}_s\|_{L_2(\partial\Omega_e)}^2 + \|c_{s_R} h_s \boldsymbol{R} \cdot \boldsymbol{e}_s\|_{L_2(\Omega_e)}^2 \tag{29}$$

$$\eta_{r_t}^2 = \|\tfrac{1}{2} c_{t_J} h_t^{1/2} \boldsymbol{J} \cdot \bar{\boldsymbol{e}}_t\|_{L_2(\partial\Omega_e)}^2 + \|c_{t_R} h_t \boldsymbol{R} \cdot \boldsymbol{e}_t\|_{L_2(\Omega_e)}^2. \tag{30}$$

4.3 Scaled anisotropic residual error indicator

For thin structures (plates or shells) the important local effects (e.g. shear stresses τ_{xz}, τ_{yz} and normal stress σ_z) have small influence in the energy norm. From an engineering point of view these small effects are important. To take care for those 3D – effects, the error indicator is augmented with an a priori scaling factor α.

$$\tilde{\eta}_e^2 = \eta_{r_e}^2 + \eta_{s_e}^2 + \alpha \cdot \eta_{t_e}^2 \tag{31}$$

The scaling factor can be chosen as

---

[1] The authors thank Professor E. Rank, Dortmund, for fruitful discussions on this topic

$$\begin{aligned}
\alpha &= 1 &&\text{for 3D--problems} \\
\alpha &= 0 &&\text{for 2D--problems for thin plates} \\
\alpha &= 10 \cdots 100 &&\text{for thin plates with composite materials} \\
\alpha &= 10 \cdots 25 &&\text{for state of stress of a plate at a column}
\end{aligned}$$

where $\sigma_z, \tau_{zx}, \tau_{zy}$ are important.

4.4 Elements for dimensional adaptivity

For adaptive mesh generation we introduce 2D and 3D elements represented by 8 corners, i.e. by volume elements. For 2D–shell elements the local stiffness matrix is calculated with selected reduced integration of the transverse shear energy (degenerated shell element) where the material tensor is given in a layered plane stress fromulation. The integration in thickness direction is numerical, because then the polynomial degree of the shape functions can be chosen different in each direction. In the following example we use degenerated shell elements with linear shape functions in each direction of the 2D–domain. Cubic shape functions are implemented in each direction of the 3D–domain and fully integrated with the 3D material tensor. For transition zones from 2D– to 3D–domains we use a degenerated shell element with quadratic shape functions in the middle surface of the shell and a cubic shape function in thickness direction in order to avoid a locking effect by changing the kinematic conditions and such to improve the local covergence.

5 Examples

In this section we present two examples illustrating the effectiveness of the present refinement strategies.

First of all, we treat the postbuckling analysis (branch–switching) of a simply supported cylinder. Adaptive meshes for the secondary equilibrium path give a deeper understanding of the buckling modes of those structures for practical engineering problems.

Fig. 2a shows an axially loaded cylinder with vanishing lateral displacements on both edges. A quarter in the circumferential direction and a half of the length is considered. Only even–numbered deformation patterns in circumferential direction are admitted. In general, the lowest bifurcation load λ_{krit} cannot be obtained by this restricted modelling; this needs the computation of the the whole system. Load factor λ versus vertical deflection U_v for system 1 is

204

presented in Fig. 2b. At equilibriuum point P_2 in the postcritical branch mesh adaptation and the real deformation is shown in Fig. 3 for the 2. adaptive mesh (3928 dof). The influence of the of the eigenvector with six waves in circumferential direction and three halfwaves in axial direction is still visible. The bifurcation mode of system 2 is demonstrated in Fig. 4. Adaptivity is performed near the bifurcation point $\lambda_{krit} = 1.09$, and Fig.4 shows the second adaptive mesh (8995 dof) with local buckles near the edges.

For h–d–adaptivity a number of domains of edge–supported haunched plates is investigated in one domain in the framework of geometrical linear theory. In Fig. 5 system parameters are given, and Fig. 6 shows the scaled h–d–adapted mesh (scaling factor is 10.0), after 10 steps of h–d–adaptivity. The subdomain in the vicinity of the support is shown in Fig. 7. One recognizes 3D–elements near the support with adjacent 2D–shell elements. If the scaling factor is not big enough the 3D–elements are generated too late in the refining process. The shear stress τ_{xz}, calculated by many refinements of the 2D–model (Fig. 8) using h–adaptivity, is compared with the results of the 3D–model (Fig. 9), using scaled h–d–adaptivity. The 2D–model doesn't yield changing shear stresses in thickness direction (Fig. 8). With the scaled h–d–adaptivity we get asymptotic convergence of the solution and the complete model.

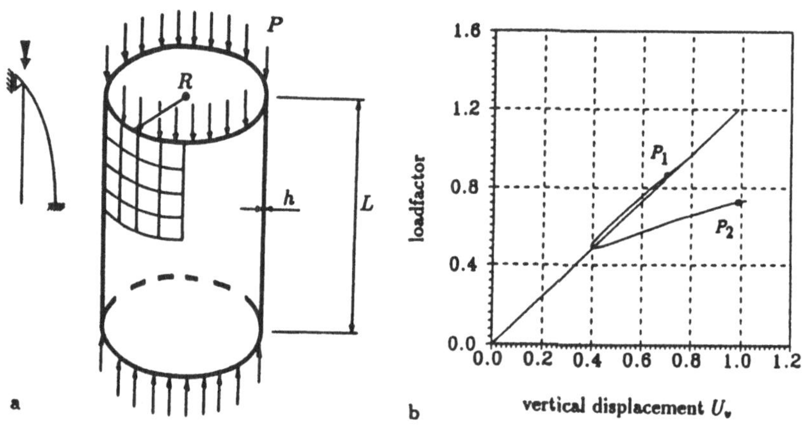

Fig. 2: **a** Cylindrical shell, geometry and loading for system 1: $R = 100m$, $L = 140m$, $h = 2m$, $E = 3.0\ 10^4 Pa$, $\nu = 0$, $P = \lambda P_{ref}$, $P_{ref} = 693.2\frac{N}{m}$; **b** load deflection curve for vertical displacement U_v

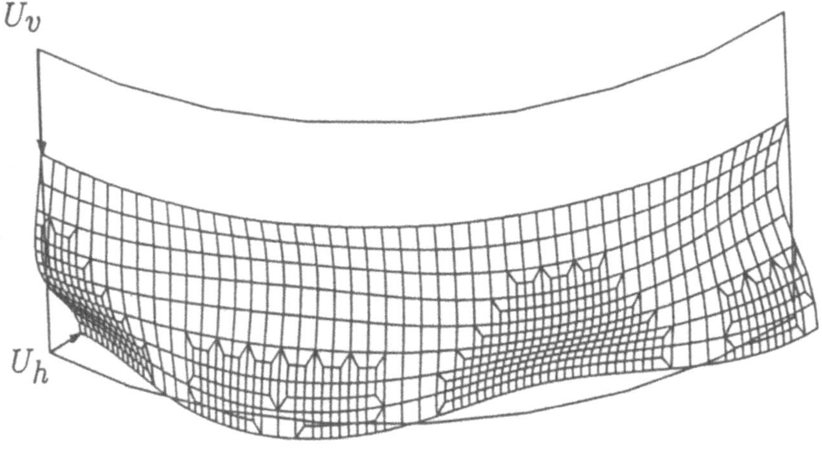

Fig. 3: Postcritical adaptive mesh and deformations for system 1 at point P_2 on the load deflection curve, see Fig. 2b.

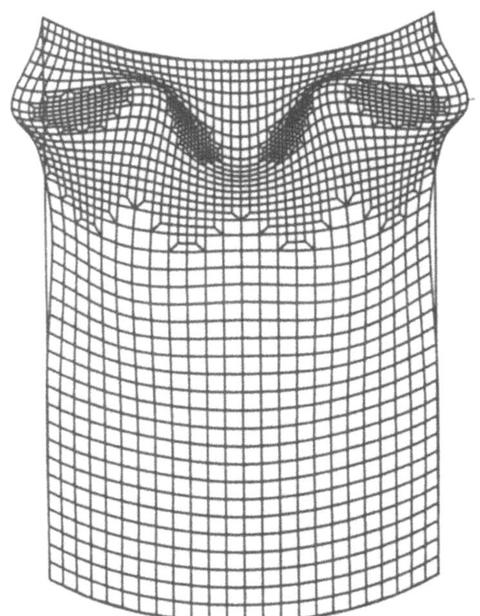

Fig. 4: Adaptive refinement for system 2 near the bifurcation point, $\lambda_{krit} = 1.09$, 2. adaptive mesh with 8995 dof; system 2 is given as follows: $R = 100mm$, $L = 500mm$, $h = 1mm$, $E = 2.1\,10^5 MPa$, $\nu = 0.3$, $P = \lambda P_{ref}$, $P_{ref} = 1000\frac{N}{mm}$, boundary conditions are the same as in system 1, Fig. 2a.

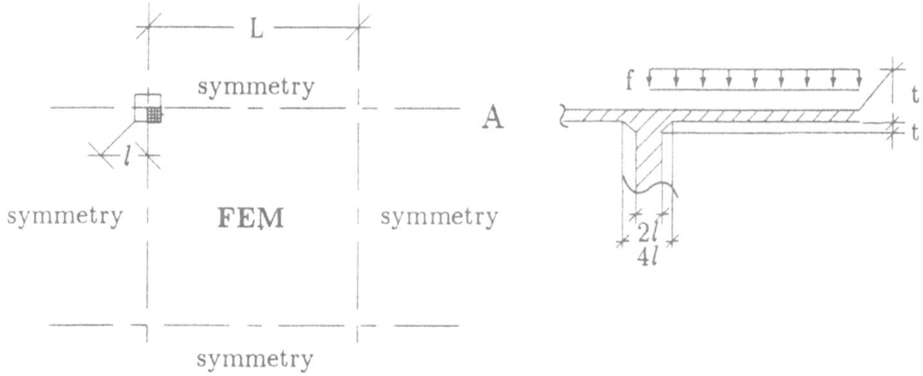

Fig. 5 A Quarter of a continuous plate system, haunched at supports, system data: L = 10 m, t = 20 cm, l =30 cm, E = 20 000 MPa , F = 100 KN, $\nu = 0.2$

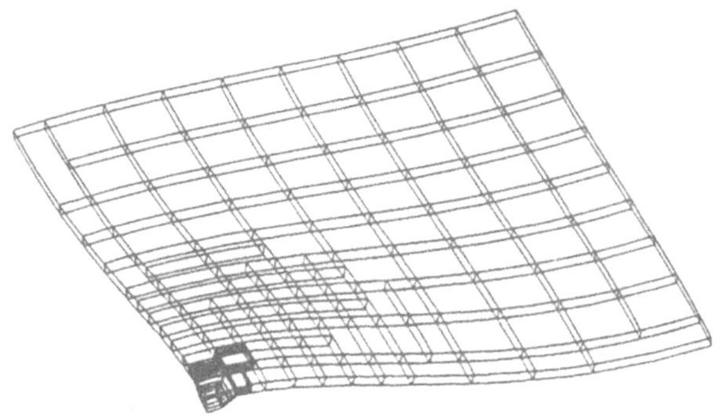

Fig. 6: Enlarged deflection of a quarter of the plate field with scaled h–d–adaptivity especially near the support.

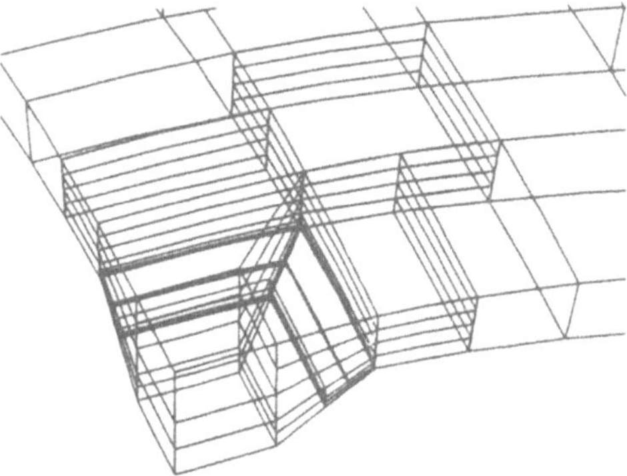

Fig. 7: Detail of Fig. 6 of the haunched plate near the support.

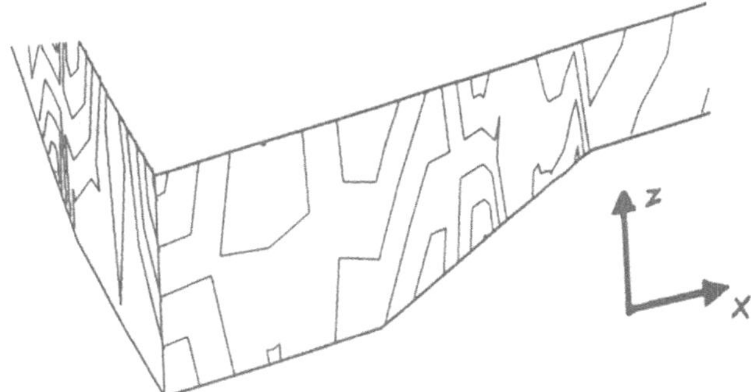

Fig. 8: Isolines of the transverse shear stress τ_{xz} near the support from 2D–plate model

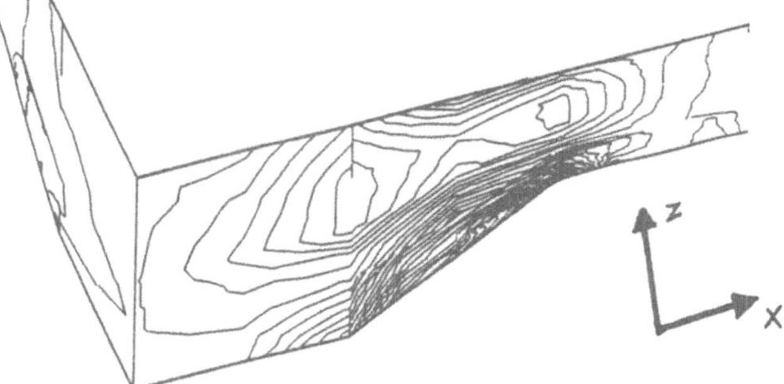

Fig. 9: Isolines of the transverse shear stress τ_{xz} near the support from complete 3D–plate model

6 Conclusions

An advanced refinement strategy is presented for geometrically non–linear plate and shell problems. The known concept of adaptivity is theoretically expanded in order to get deeper insight into bifurcation problems and related questions. One main result of this paper is that the used concept of adaptivity is not possible directly at bifurcation points and that asymptotic convergence cannot be obtained for locking dominated problems. The robustness of the chosen mechanical and numerical model has to be investigated separately, what is evident.

The whole strategy is completed by the dimensional adaptivity of disturbed 2D – subdomains which allows a reliable and flexible modelling and computation. The computer code was developed within our system INA–SP; it is a flexible tool for a rather general class of non–linear problems.

208

References

[1] Babuška I. & Miller A.: A feedback finite element method with a posteriori error estimates: Part I. The finite element mothod and some properties of the a posteriori estimator, Comp. Meth. Appl. Mech. Engng. 61 (1987).

[2] Babuška I. & Rheinhodt W.C.: Error estimates for adaptive finite element computations, SIAM Journal on Numerical Analysis 15 (1978) 736-754.

[3] Johnson, C. and Hansbo, P.: Adaptive finite element methods in computational mechanics, Comp. Meth. Appl. Mech. Engng. 101 (1992) 143-181.

[4] Gruttmann F., Stein E. & Wriggers P.: Theory and Numerics of Thin Elastic Shells with Finite Rotations, Ingenieur Archiv 59 (1989) 54-67.

[5] Rheinboldt W.C.: Error estimates for nonlinear finite element computations. Computers and Structures Vol. 20 (1985) pp. 91–98.

[6] Simo J.C., Fox D.D. & Rifai M.S.: On a Stress Resultant Geometrically Exact Shell Model, Part III: Computational Aspects of the Nonlinear Theory, Comp. Meth. Appl. Mech. Engng. 79 (1990) 21-70.

[7] Stein E., Rust W. & Ohnimus S.: h- and d-adaptive FE element methods for two dimensional structural problems including post-buckling of shells, In: Proceedings of the Second Workshop on Reliability in Computational Mechanics (Krakov, 1991), Comp. Meth. Appl. Mech. Engng. 101 (1992) 315-354.

[8] Stein E. & Ohnimus S.: Concept and realisation of integrated adaptive finite element methods in solid- and structural-mechanics in: Proceedings of the First European Conference on Numerical Methods in Engeneering (7 - 11 Sept. 1992, Brussels, Belgium), Int. J. Num. Meth. Eng. (1992) 163-170.

[9] Stein, E. & Carstensen, C. & Seifert, B. & Ohnimus, S.: Adaptive finite element analysis of geometrically non–linear plates and shells, especially Buckling, Int. J. Num. Meth. Eng., in press.

[10] Verfürth R.: A posteriori error estimates for non-linear problems. Finite element discretizations of elliptic equations (preprint 1993).

[11] Zienkiewicz O.C. & Zhu J.Z.: A simple error estimator and adaptive procedure for practical engineering analysis, Int. J. Num. Meth. Eng. 24 (1987) 337-357.

Erwin Stein, Bengt Seifert, Stephan Ohnimus
University of Hannover, Germany
Institute of Structural and Computational Mechanics (IBNM)
Callinstra e 9A
30167 Hannover
e-mail: stein@leibniz.ibnm.uni-hannover.d400.de

MULTIGRID METHOD ON UNSTRUCTURED MESHES

Petr Vaněk

1 Introduction

In this paper a coarsening strategy appropriate for unstructured meshes will be described. The main idea of the presented technique consists in combining of the unknowns aggregation and smoothing. The advantage of the unknowns aggregation is that it can be easily done for very general meshes. However the resulting coarse spaces contain high energy functions what results in poor convergence properties. The role of smoothers used in the coarsening is to suppress the energy of the coarse space functions. This idea was described by the author in [5]. The AMG-like solver based on it was suggested and tested in [6]. The two-level preconditioner with a small coarse grid based on the multiple usage of a smoother has been suggested and analyzed in [7]. In the algorithm presented in this paper the unknowns aggregation is done using hierarchy of decompositions of the domain. Such approach enables to manage unstructured meshes and moreover the pattern of the resulting coarse grid matrices is regular and follows the nine-point stencil. This paper contains the description of the algorithm only, both convergence analysis and numerical experiments can be found in [2]. In the end of this contribution the convergence theorem from [2] is cited without proving.

2 The Problem Formulation and Multigrid Algorithm

Let us consider, for simplicity, the following Dirichlet problem. For a given $f \in L^2(\Omega)$ find $u \in H_0^1(\Omega)$ such that

$$(2.1) \qquad A(u,v) = (f,v) \ \text{ for all } \ v \in H_0^1(\Omega),$$

where $A(.,.)$ is a symmetric, V-elliptic continuous bilinear form on $H_0^1(\Omega)$, $(.,.)$ is the L^2-scalar product, $\Omega \subset {\rm I\!R}^2$ a polygonal (possibly nonconvex)

210

domain.

We assume a quasiuniform triangulation $\tau_h = \{T_j\}$ is given on Ω. The corresponding piecewise linear finite element subspace of $H_0^1(\Omega)$ we denote by V_h and the meshsize by h_1.

For the purpose of a definition of the multigrid algorithm let us set $M_1 = V_h$. We need a hierarchy of coarse spaces

$$M_1 \supset M_2 \supset \ldots \supset M_L .$$

In addition, let $(.,.)_l$ be scalar products on M_l, $||.||_l = (.,.)_l^{1/2}$ the corresponding norms. We will describe the multigrid algorithm for the solution of the finitedimensional problem: for a given $f \in M_1$ find $u \in M_1$ such that

$$(2.2) \qquad\qquad A(u, \Phi) = (f, \Phi)_1 \text{ for all } \Phi \in M_1.$$

To define the multigrid algorithm, we will define auxiliary operators $A_l : M_l \to M_l$, $l = 1, \ldots, L$ by

$$A(w, \Phi) = (A_l w, \Phi)_l \quad \text{for all} \quad w, \Phi \in M_l.$$

Also define restrictions $P_l^0 : M_1 \to M_l$, $l = 2, \ldots, L$ by

$$(P_l^0 w, \Psi)_l = (w, \Psi)_{l-1} \quad \text{for all} \quad w \in M_{l-1}, \Psi \in M_l.$$

To define the smoothing process we require linear operators $R_l : M_l \to M_l$, $l = 1, \ldots, L - 1$. We assume that that R_l is positive semidefinite and set

$$K_l = I - R_l A_l .$$

We define the multigrid operator $B_1 : M_1 \to M_1$ by induction. Set $B_L = A_L^{-1}$. Assume that B_{l+1} has been defined and define $B_l g$ for $g \in M_l$ as follows.

1. set $x = 0$,

2. for $i = 1, \ldots, m(l)$
$$x \leftarrow x + R_l(g - A_l x) ,$$

3. set $x \leftarrow x + q$, where q is defined by
$$q = 0 ,$$
for $i = 1, \ldots, p$
$$q \leftarrow q + B_{l+1}[P_{l+1}^0(g - A_l x) - A_{l+1} q] ,$$

4. for $i = 1, \ldots, m(l)$
$$x \leftarrow x + R_l(g - A_l x) \, ,$$

5. set $B_l g = x$.

In this algorithm, $m(l)$ is a positive integer which may vary from level to level and determines the number of smoothing iterations on that level. The number p is a positive integer as well. Cases $p = 1$, $p = 2$ correspond to the symmetric V and W cycles of multigrid, respectively.

3 Construction of Coarse Spaces

In this section an algorithm of construction of coarse spaces $M_2, \ldots, M_L$ will be described. Coarsening will be done in two main steps. In the first one, hierarchy of auxiliary prolongation operators p_l will be constructed using the technique of aggregation of unknowns. Sets of aggregated nodes will be defined with help of a decomposition of Ω.

In the second step we will define smoothers successively and using them we will improve the operators p_l and construct coarse level matrices. Roughly speaking, the construction will follow this scheme:

INPUT: auxiliary prolongation operators $p_1, \ldots, p_{L-1}$,
($p_l : (l+1)$–th level $\rightarrow$ l–th level), $\widehat{A}_1$ the stiffness matrix.

OUTPUT: improved prolongation operators, hierarchy of coarse level matrices for $l = 1, \ldots, L - 1$

1. define a smoother $\widehat{S}_l$ using $\widehat{A}_l$

2. compute the improved prolongation operator $\widehat{p}_l = \widehat{S}_l p_l$

3. compute the matrix on the $(l+1)$–th level $\widehat{A}_{l+1} = \widehat{p}_l^T \widehat{A}_l \widehat{p}_l$.

Due to the complexity of the algorithm we want the coarse level matrices to keep away from fill-in. Because of it in the first step we will use a first degree polynomial of $\widehat{A}_l$ as a smoother. Moreover the auxiliary prolongation operators will be constructed so that the usage of smoothers in the smoothing process will result in no increase of fill-in. More precisely, auxiliary prolongation operators will be defined so that the matrices $p_l^T A_l p_l$ and $\widehat{p}_l^T \widehat{A}_l \widehat{p}_l$ will have the same fill-in and the pattern following the nine-point stencil. In

212

the other words, regardless of the fact that the finite element mesh may be unstructured the structure of coarse level matrices will be regular.

Let us start with construction of the auxiliary prolongation operators. The domain Ω will be covered by an auxiliary regular square grid of a grid size h_2. This auxiliary grid divides Ω into a system of subdomains Ω_i^1.

Later on we will formulate the requirement on the choice of h_2 which will quarantee the sparseness and the regular pattern of the coarse level matrices mentioned above. Here we suppose only that in each subdomain Ω_i^1 there is at least one triangle of τ_h.

Let us denote by $\widehat{D}^1$ the set of all subdomains Ω_i^1, and let us set

$$D^1 = \{\Omega_i^1 \in \widehat{D}^1 \;:\; \partial\Omega_i^1 \cap \partial\Omega \;=\; \emptyset \;\} \;.$$

Each subdomain $\Omega_i^1 \in D^1$ will be represented by one degree of freedom on the second level. This degree of freedom will be associated with a node v_i^2 of the triangulation τ_h contained in Ω_i^1 (so called c-point)
. Now we will describe the way how to choose the c-point v_i^2. Let us denote by Ω_i^h the union of all triangles from τ_h which are contained in Ω_i^1. The c-point will be chosen approximately in "the middle" of Ω_i^1, more precisely, v_i^2 will be chosen so that the distance of v_i^2 and $\partial\Omega_i^h$ measured by the number of edges will be maximal. Let $N^1 = \{v_i^1\}$ denote the set of all unconstrained nodes in τ_h, $N^2 = \{v_i^2\}$ the set of all c-points on the second level. Further we will need n_i - the number of elements of N^i. To each $\Omega_i^1 \in D^1$ we will assign an index set

$$C_i^1 = \{j : v_j^1 \in \Omega_i^1\}.$$

Let us set

$$\alpha_1 = (max|C_i^1|)^{-1/2}.$$

An auxiliary prolongation operator $p_1 : \mathbb{R}^{n_2} \longrightarrow \mathbb{R}^{n_1}$ will be defined

$$(p_1)_{ij} = \left\{ \begin{array}{l} \alpha_1 \;\; \text{if} \;\; i \in C_j^1 \\ 0 \;\; \text{otherwise} \end{array} \right. .$$

In fact, the value from the c-point v_i^2 (multiplied by a scaling factor α_1) is loaded onto all nodes from τ_h contained in $\Omega_i^1 \in D^1$.

Convenience of scaling by α_1 will become evident in the convergence analysis, see [2], for the scaling guarantees the inequality

$$(p_1 x)^T p_1 x \leq x^T x \;\; \text{for all} \;\; x \in \mathbb{R}^{n_2}.$$

Further coarsening (i.e. the aggregation of nodes from N^2 and choice of c-points $N^3 \subset N^2$) will be done as follows. At first, we will construct an auxiliary regular square grid of the grid-size $h_3 = 3h_2$ aligned with the previous square grid. Again, this grid divides Ω into subdomains Ω_i^2. Let us denote by $D^2 = \{\Omega_i^2 : \partial\Omega_i^2 \cap \partial\Omega = \emptyset\}$. Each subdomain $\Omega_i^2 \in D^2$ consists of nine subdomains from D^1. The c-point v_i^3 of $\Omega_i^2 \in D^2$ will be the c-point v_k^2 of Ω_k^1 which is in the middle of Ω_i^2 .

An auxiliary prolongation operator p_2 will be defined analogously to p_1, i.e.

$$(p_2)_{ij} = \begin{cases} \alpha_2 & \text{if } v_i^3 \in \Omega_j^2 \\ 0 & \text{otherwise} \end{cases}.$$

As every Ω_i^2 contains exactly nine nodes of N^2, we set $\alpha_2 = \frac{1}{3}$. Obviously

$$(p_2 x)^T p_2 x = x^T x \quad \text{for all } x \in \mathbb{R}^{n_3}, \, n_3 = |N^3|$$

We will go on with the coarsening using the same procedure (i.e., constructing $p_3, p_4, \ldots, N^4, N^5, \ldots$) until we reach the L-th level for which $n_L = |N_L|$ is small enough so that the corresponding system of linear algebraic equations can be solved using direct methods.

It is not difficult to see that

$$(3.1) \qquad\qquad \alpha_1 \cdot \alpha_2 \cdots \alpha_k \approx 3^{l-1}$$

and that there exists a constant $C > 0$ depending on $\frac{h_2}{h_1}$ only such that

$$C \, x^T x \leq (p_1 p_2 \ldots p_{l-1} x)^T (p_1 p_2 \ldots p_{l-1}) \leq x^T x \quad \text{for every } x \in \mathbb{R}^{n_l}.$$

The first step of our algorithm has been completed. To accomplish the second step, we will describe more precisely the scheme sketched in the beginning of the section. Let us define sequences $\{\widehat{A}_l\}_{l=1}^L$, $\{\widehat{S}_l\}_{l=1}^{L-1}$, $\{\bar{\lambda}_l\}_{l=1}^{L-1}$, $\{\tilde{p}_l\}_{l=1}^{L-1}$ by

$$(3.2) \qquad \begin{aligned} \bar{\lambda}_l &= \begin{cases} \bar{\varrho}(\widehat{A}_1) & \text{for } l = 1 \\ \min\{\tfrac{1}{9}\bar{\lambda}_{l-1}, \, \bar{\varrho}(\widehat{A}_l)\} & \text{for } l > 1 \end{cases}, \\ \widehat{S}_l &= I - \frac{\omega}{\bar{\lambda}_l}\widehat{A}_l, \\ \tilde{p}_l &= \widehat{S}_l p_l, \\ \widehat{A}_{l+1} &= \tilde{p}_l^T \widehat{A}_l \tilde{p}_l, \end{aligned}$$

214

where $\bar{\varrho}(\widehat{A}_l)$ is a majorant of $\varrho(\widehat{A}_l)$ satisfying

(3.3) $$\varrho(\widehat{A}_l) \leq \bar{\varrho}(\widehat{A}_l) \leq c\varrho(\widehat{A}_l)$$

and

$$\omega = \frac{4}{3}.$$

Let us note that (3.2) immediately implies

$$\bar{\lambda}_{l+1} \leq \frac{1}{9}\bar{\lambda}_l .$$

In [2] will be shown that

$$\varrho(\widehat{A}_l) \leq \bar{\lambda}_l \leq c\varrho(\widehat{A}_l).$$

We are now ready to define coarse spaces $M_2, \ldots, M_L$ and the scalar products associated with them.

Let $I_1 : \mathbb{R}^{n_1} \to V_h$ denote the interpolation operator defined by

$$I_1 x = \sum_{i=1}^{n_1} x_i \psi_i^1,$$

where $\{\psi_i^1\}_{i=1}^{n_1}$ is a standard nodal basis of $V_h \subset H_0^1(\Omega)$ corresponding to the unconstrained nodes.

Let us define operators $S_l^1 : \mathbb{R}^{n_l} \to \mathbb{R}^{n_1}$ by

$$S_l^1 = \begin{cases} \text{identity on } \mathbb{R}^{n_1} & \text{for } l = 1 , \\ \tilde{p}_1 \tilde{p}_2 \ldots \tilde{p}_{l-1} & \text{for } l = 2, \ldots, L. \end{cases}$$

Definition 3.1 *For $l = 2, \ldots, L$ define $M_l = \mathrm{Range}(I_1 S_l^1)$.*

Note that any $x \in M_l$ can be written in the form

$$x = I_1 S_l^1 \widehat{x}, \ \widehat{x} \in \mathbb{R}^{n_l}.$$

Definition 3.2 *For arbitrary $x, y \in M_l$, $l = 1, \ldots, L$, define the scalar product*

(3.4) $$(x, y)_l = \widehat{x}^T \widehat{y},$$

where $x = I_1 S_l^1 \widehat{x}$, $y = I_1 S_l^1 \widehat{y}$.

In [2] the following equivalence is proved:

$$(3.5) \qquad C_1||x||_{0,\Omega} \leq h_1||x||_{0,\Omega} \leq C_2||x||_{0,\Omega} \ \text{ for all } \ x \in M_l,$$

where C_1, $C_2 > 0$ are constants independent of l, h_1. This equivalence guarantees that the scalar products in Definition 3.2 are well defined.

At the end of the section, we formulate the requirement on h_2 which will guarantee sparsity and the regular pattern of the coarse-level matrices . To this end, we introduce additional notation.

Let $\Omega_i^1 \in \widehat{D}^1$. Denote

$$A_i = \{T_j \subset \tau_h : T_j \cap \Omega_i^1 \neq \emptyset\},$$

i.e., A_1 is the set of all triangles intersecting Ω_i^1,

$$\Omega(A_i) = \bigcup_{T_j \in A_i} T_j,$$

$$B_i = A_i \cup \{T_j \in \tau_h : \partial T_j \cap \partial\Omega(A_i) \neq \emptyset\},$$

$$\Omega(B_i) = \bigcup_{T_j \in B_i} T_j,$$

i.e., we add one strip of triangles to $\Omega(A_i)$. It is easy to see that $\{I_1 S_l^1 e_i^l\}_{i=1}^{n_l}$ form a basis of M_l, e_i^l being the i-th vector of the canonical basis of $\mathbb{R}^{n_l}$. Apparently for $\Omega_i^1 \in D^1$

$$\Omega(A_i) = \mathrm{supp}(I_1 p_1 e_i^2),$$

$$\Omega(B_i) \supset \mathrm{supp}(I_1 \widehat{S}_1 p_1 e_i^2).$$

ASSUMPTION 1   *1. For any two subdomains Ω_i^1, $\Omega_j^1 \in D^1$ with a common side there exists a dashed line l_{ij} connecting c-points v_i^2 and v_j^i satisfying: if for $\Omega_k^1 \in \widehat{D}^1$: $l_{ij} \cap \mathrm{int}\Omega(B_k) \neq \emptyset$ then $\Omega_k^1 = \Omega_i^1$ or $\Omega_k^1 = \Omega_j^1$.*

2. For any subdomain $\Omega_i^1 \in D^1$ with at least one neighbour from $\widehat{D}^1 \backslash D^1$ (i.e. at least one neighbour of Ω_i^1 is a boundary subdomain) there exists a dashed line l_i from the c-point v_i^2 to some node $w \in \partial\Omega$ satisfying: if for $\Omega_k^1 \in \widehat{D}^1$: $l_i \cap \mathrm{int}\Omega(B_k) \neq \emptyset$ then $\Omega_k^1 = \Omega_i^1$ or $\Omega_k^1 \in \widehat{D}^1 \backslash D^1$ and Ω_k^1 shares the side with Ω_i^1 .

Remark 3.1 As it was mentioned above, domains $\{\Omega(B_i) : \Omega_i^1 \in D^1\}$ will play the role of supports of basis functions of the space M_2. Thus 1 guarantees that the restriction of a function $u \in M_2$ to l_{ij} is spanned by i-th and j-th basis functions only. 2 can be interpreted similarly.

Remark 3.2 It is easy to see that Assumption 1 implies the following:

$$\partial\Omega_i^1 \cap \partial\Omega_j^1 = \emptyset \Longrightarrow \mathrm{int}\Omega(B_i) \cap \mathrm{int}\Omega(B_j) = \emptyset.$$

This property will enable us to show that the stencil of coarse-level matrices will follow the 9-point scheme.

4   Implementation

This section addresses the issues of implementation of the multigrid algorithm with the coarse spaces as described in Section 3. We will start with the following lemma summarizing several simple properties of spaces $H_1 \supset \ldots \supset H_L$ and operators $\{A_l\}_{l=1}^L$. The lemma is not difficult to prove by verifying the statements in the presented order.

Lemma 4.1 *Let the symbol $*$ denote the adjoint operator with respect to the Euclidean scalar product in $\mathbb{R}^{n_l}$ and $(.,.)_l$. The following relations hold*

$$(4.1) \qquad\qquad (I_1 S_l^1)^* = (I_1 S_l^1)^{-1},$$

$$(4.2) \qquad\qquad \widehat{A}_l = (I_1 S_l^1)^{-1} A_l (I_1 S_l^1),$$

$$(4.3) \qquad\qquad p_{l-1}^T \widehat{S}_{l-1} = (I_1 S_l^1)^{-1} P_l^0 (I_1 S_{l-1}^1)$$

for $l = 1, \ldots, L$.

The equality (4.2) states that the matrix $\widehat{A}_l$ is a representation of the operator A_l with respect to the basis $I_1 S_l^1 \{e_i^l\}$ of H_l, $\{e_i^l\}$ being the canonical basis of $\mathbb{R}^{n_l}$. Similarly (4.3) states that $p_{l-1}^T \widehat{S}_{l-1}$ is the representation of $P_l^0 : H_{l-1} \to H_l$. It is easy to see that $\widehat{S}_l p_l$ represents the identity $H_{l+1} \to H_l$. Thus, in order to implement the algorithm described in Section 2, one may simply rewrite it in the "vector form". We will use the smoother $R_l = \frac{\tau}{\bar{\lambda}_l}$, where $\tau \in (0, 1)$ and $\bar{\lambda}_l$ was introduced in (3.2).

Let $\widehat{g}_1 \in \mathbb{R}^{n_1}$ denote the right-hand side. We may rewrite the abstract algorithm as follows:

$$
\boxed{
\begin{aligned}
&\textit{if } l = L, \textit{ set } \widehat{B}_l = \widehat{A}_L^{-1} \\
&\textit{else} \\
&\qquad \textit{set } \widehat{x}_l = 0,\ \widehat{x}_l \in \mathbb{R}^{n_l}, \\
&\qquad \textit{for } i = 1, \ldots, m(l) \\
&\qquad\qquad \widehat{x}_l \leftarrow \widehat{x}_l + \frac{\tau}{\lambda_l}(\widehat{g}_l - \widehat{A}_l\widehat{x}_l), \\
&\qquad \textit{set } \widehat{x}_l \leftarrow \widehat{x}_l + \widehat{S}_l p_l \widehat{q}_{l+1}, \textit{ where } \widehat{q}_{l+1} \in \mathbb{R}^{n_{l+1}} \textit{ is defined by:} \\
&\qquad \widehat{q}_{l+1} = 0, \\
&\qquad \textit{for } i = 1, \ldots, m(l) \\
&\qquad\qquad \widehat{q}_{l+1} \leftarrow \widehat{q}_{l+1} + \widehat{B}_{l+1}[p_l^T \widehat{S}_l(\widehat{g}_l - \widehat{A}_l\widehat{x}_l) - \widehat{A}_{l+1}\widehat{q}_{l+1}], \\
&\qquad \textit{for } i = 1, \ldots, m(l) \\
&\qquad\qquad \widehat{x}_l \leftarrow \widehat{x}_l + \frac{\tau}{\lambda_l}(\widehat{g}_l - \widehat{A}_l\widehat{x}_l), \\
&\qquad \textit{set } \widehat{B}_l\widehat{g}_l = \widehat{x}_l.
\end{aligned}
}
$$

$$\textbf{Algorithm 1}$$

The algorithm $\widehat{B}_1 : \widehat{g}_1 \to \widehat{x}_1$ returns the approximation $\widehat{x}_1$ of the solution of the system of linear algebraic equations $\widehat{A}_1\widehat{x} = \widehat{g}_1$. The iterative process will be performed in this manner:

$$
\widehat{x}_1 \leftarrow \widehat{x}_1 + \widehat{B}_1(\widehat{A}_1\widehat{x}_1 - \widehat{g}_1).
$$

The following convergence theorem can be found in [2].

Theorem 1 *Under Assumption 1, the rate of convergence of Algorithm 1 with respect to the energy norm is bounded by $1 - \frac{C}{L}$, where $C > 0$ is a constant independent on h_1 and L denotes the number of levels. Moreover, there are maximally nine nonezero entries in one row of each coarse grid matrix.*

References

[1] Bramble J.H., Pasciak J.E., Wang J., Xu J. : *Convergence Estimates for Multigrid Algorithm without Regularity Assumptions*, Math.Comp. 57(1991)

[2] Mandel J. ,Vaněk P., Brezina M. : *Multigrid on Unstructured Meshes*, to appear.

[3] Mandel J. : *Balancing Domain Decomposition*, Communications in Numerical Methods in Engineering, Vol.9 (1993).

[4] Hackbusch W. : *Multi–Grid Methods and Applications*, Springer–Verlag, 1985

[5] Vaněk P. : *Acceleration of a Two–level Algorithm by Smoothing Transfer Operators*, Appl.of Math.37(1992), No.4.

[6] Vaněk P. : *Fast Multigrid Solver* , to appear in Appl. of Math .

[7] Vaněk P., Křížková J. : *Two-level Preconditioner with Small Coarse Grid Appropriate for Unstructured Meshes*, to appear.

Author's address: Dep.of Math.,University of West Bohemia, 306 14 Plzeň, Czech Republic
e-mail: pvanek@scylla.zcu.cz

List of Participants

Alber, Hans-Dieter, Darmstadt
Ali Mehmeti, Felix, Darmstadt
Antonevich, Anatoliĭ, Minsk
Apel, Thomas, Chemnitz
Assaf, Mohammed, Chemnitz
Babovsky, Hans, Heidelberg
Bach, Michael, Stuttgart
Bagirov, Lev, Moscow
Beckert, Herbert, Markkleeberg
Beer, Klaus, Chemnitz
Benda, Josef, Prague
Blaheta, Radim, Ostrava
Bollhöfer, Matthias, Chemnitz
Bärwolf, Günter, Berlin
Böttcher, Albrecht, Chemnitz
Carstensen, Carsten, Hannover
Chkadua, Otari, Tbilisi
Chleboun, Jan, Prague
Costabel, Martin, Rennes
Daniels, Helmut, Heidelberg
Datschew, Geno, Sofia
Dempe, Stephan, Chemnitz
Didenko, Viktor, Odessa
Dresig, Hans, Chemnitz
Duduchava, Roland, Tbilisi
Dybin, Vladimir, Rostov on Don
Dümmel, Siegfried, Chemnitz
Eck, Christof, Stuttgart
El-Dessouki, Nadja, Stuttgart
Elschner, Johannes, Berlin
Eppler, Karsten, Chemnitz
Feistauer, Miloslav, Prague
Felgenhauer, Andreas, Freiberg
Fernandez, Luis Alberto, Santander
Filinov, Vladimir, Moscow
Finck, Tilo, Chemnitz
Fischer, Andreas, Berlin

Fleischer, Wolfgang, Chemnitz
Fuhrmann, Jürgen, Berlin
Gawinecki, Jerzy, Warsaw
Gebel, Michael, Freiberg
Globisch, Gerhard, Chemnitz
Goldberg, Helmuth, Chemnitz
Goldstein, Robert, Moscow
Gottlieb, Johannes, Karlsruhe
Grudskiĭ, Sergeĭ, Rostov on Don
Gründemann, Heinz, Mittweida
Gürlebeck, Klaus, Chemnitz
Haase, Gundolf, Chemnitz
Hackbusch, Wolfgang, Kiel
Haftmann, Rolf, Chemnitz
Hagen, Roland, Chemnitz
Hartmann, Thomas, Hannover
Hausding, Reiner, Chemnitz
He, Chunyang, Chemnitz
Hebeker, Friedrich-Karl, Heidelberg
Heinrich, Bernd, Chemnitz
Heise, Bettina, Chemnitz
Heise, Bodo, Chemnitz
Hengst, Sabine, Berlin
Hermann, Peter, Aachen
Hoffmann, Karl-Heinz, München
Hofmann, Bernd, Chemnitz
Holenda, Jiři, Plzeň
Hommel, Angela, Chemnitz
Hommel, Thomas, Chemnitz
Hsiao, George, Delaware
Jentsch, Lothar, Chemnitz
Jung, Michael, Chemnitz
Junghanns, Peter, Chemnitz
Kapurkin, Andreĭ, Magdeburg
Kisil, Vladimir, Odessa
Knabner, Peter, Berlin
Kočvara, Michal, Prague

Kondratev, Vladimir, Moscow
Korneev, Vadim Glebovich, St Petersburg
Kozel, Karel, Prague
Kravchenko, Vladislav, Odessa
Křižková, Jitka, Plzeň
Kunert, Gerd, Chemnitz
Kunoth, Angela, Berlin
Lang, Wilgard, Chemnitz
Langer, Ulrich, Linz
Langmach, Hartmut, Fredersdorf
Lawin, Roger, Chemnitz
Lebre, Amarino, Lisbon
Lezius, Ralf, Chemnitz
Liebold, Gerhard, Chemnitz
Lowke, Friedmar, Chemnitz
Lube, Gert, Magdeburg
Luderer, Bernd, Chemnitz
Martensen, Erich, Karlsruhe
Maslovskaya, Larisa V., Odessa
Mauermann, Werner, Chemnitz
Mehrmann, Volker, Chemnitz
Meiners, Wolfgang, Paderborn
Meister, Erhard, Darmstadt
Meyer, Arnd, Chemnitz
Mikhailets, Vladimir, Kiev
Mirschinka, Dirk, Chemnitz
Mottelet, Stephane, Compiègne
Männikkö, Timo, Jyväskylä
Natroshvili, David, Tbilisi
Nečas, Jindřich, Prague
Nepomnyashchikh, Sergeĭ, Novosibirsk
Neundorf, Werner, Ilmenau
Nicaise, Serge, Valenciennes
Oestreich, Dieter, Dresden
Okhezin, Sergeĭ, Yekaterinburg
Pedas, Arvet, Tartu
Penzel, Frank, Darmstadt

Pester, Matthias, Chemnitz
Piskorek, Adam, Warsaw
Pobedria, Boris, Moscow
Popa, Constantin, Constanţa
Prößdorf, Siegfried, Berlin
Rannacher, Rolf, Heidelberg
Roch, Steffen, Chemnitz
Roche, Jean, Nancy
Rost, Karla, Chemnitz
Rüde, Ulrich, München
Rönsch, Wolfgang, Heidelberg
Rösch, Arnd, Chemnitz
Rösch, Pia, Chemnitz
Saranen, Jukka, Oulu
vom Scheidt, Jürgen, Chemnitz
Schell, Hans-Joachim, Chemnitz
Scherzer, Matthias, Chemnitz
Schmutzler, Bernd, Freiberg
Schulz, Jürgen, Chemnitz
Shaĭdurov, Vladimir, Krasnoyarsk
Shargorodskiĭ, Eugene, Tbilisi
Shutyaev, Victor, Moscow
Silbermann, Bernd, Chemnitz
Socolowsky, Jürgen, Halle
Steidten, Torsten, Chemnitz
Stein, Erwin, Hannover
Steinbach, Jörg, München
Steinbach, Olaf, Stuttgart
Svanadse, Merab, Tbilisi
Szöcs, Huba, Székesfehérvár
Sändig, Anna-Margarete, Stuttgart
Tautenhahn, Ulrich, Chemnitz
Tiba, Dan, München
Triebsch, Falk, Chemnitz
Tröltzsch, Fredi, Chemnitz
Uba, Peep, Tartu
Uhrlandt, Dirk, Greifswald
Vaněk, Petr, Plzeň
Weber, Beate, Chemnitz

Weber, Uwe, Chemnitz
Wegert, Elias, Freiberg
Weigand, Peter, Chemnitz
Weinelt, Wilfried, Chemnitz
Wendland, Wolfgang, Stuttgart
Windisch, Günther, Chemnitz
Winzer, Sirko, Dresden

von **Wolfersdorf**, Lothar, Freiberg
Wolska–Bochenek, Janina, Warsaw
Würker, Uwe, Chemnitz
Yserentant, Harry, Tübingen
Zalaletdinov, Roustam M., Tashkent
Żochowski, Antoni, Warsaw

Walser
Der
Goldene Schnitt

Der Goldene Schnitt tritt seit der Antike
in vielen Bereichen der Geometrie,
Architektur, Musik, Kunst sowie der
Philosophie auf, aber er erscheint auch
in neueren Gebieten der Technik und
der Fraktale. Dabei ist der Goldene
Schnitt kein isoliertes Phänomen, son-
dern in vielen Fällen das erste und so-
mit einfachste nichttriviale Beispiel im
Rahmen weiterführender Verallgemei-
nerungen.
Ziel dieses Buches ist es, einerseits
Beispiele des Goldenen Schnittes zu
besprechen, andererseits weiterführen-
de Wege aufzuzeigen.

Von **Hans Walser,**
Frauenfeld

1993. 140 Seiten.
13,7 x 20,5 cm.
Kart. DM 16,80
ÖS 131,– / SFr 16,50
ISBN 3-8154-2070-9

(Einblicke in die Wissen-
schaft – Mathematik)

Koprod. Verlag der
Fachvereine, Zürich –
B. G. Teubner, Leipzig

B. G. Teubner Verlagsgesellschaft
Stuttgart · Leipzig

Bronstein/ Semendjajew

Taschenbuch der Mathematik

Im Vorwort zur ersten deutschen Auflage, die 1958 im Verlag B. G. Teubner Leipzig erschien, heißt es zur Zielsetzung des Werkes:

Mit der Herausgabe der deutschen Übersetzung des Taschenbuches der Mathematik von Bronstein und Semendjajew hofft der Verlag, den angehenden und in der Praxis stehenden Ingenieuren und darüber hinaus auch Physikern und Mathematikern ein wirklich brauchbares Nachschlagewerk in die Hand zu geben und damit eine empfindliche Lücke in der deutschen mathematischen Literatur zu schließen. Auch als Repetitorium der Mathematik dürfte das Buch gute Dienste leisten.

Die vorliegende 25. Auflage basiert auf der 1979 völlig überarbeiteten 19. Auflage. Seine Vorzüge hat das Werk wohl am besten dadurch unter Beweis gestellt, daß seither 25 Auflagen mit über 800.000 Exemplaren erschienen sind.

Aus dem Inhalt:

Tabellen und graphische Darstellungen – Elementarmathematik – Analysis – Mengen, Relationen, Funktionen, Vektorrechnung, Differentialgeometrie, Fourierreihen, Fourierintegrale, Laplacetransformation – Wahrscheinlichkeitsrechnung und mathematische Statistik – Lineare Optimierung – Numerik

Von
Ilja N. Bronstein
und
Konstantin A. Semendjajew
Moskau

Herausgegeben von
Günter Grosche,
Viktor Ziegler und
Dorothea Ziegler, Leipzig

25. Auflage. 1991. XII, 840 Seiten mit 390 Bildern. 14,5 x 20 cm.
Geb. DM 36,–
ÖS 281,– / SFr 36,–
ISBN 3-8154-2000-8

Alleinauslieferung:
B. G. Teubner Stuttgart

B. G. Teubner Verlagsgesellschaft Stuttgart · Leipzig

Function Spaces, Differential Operators and Nonlinear Analysis

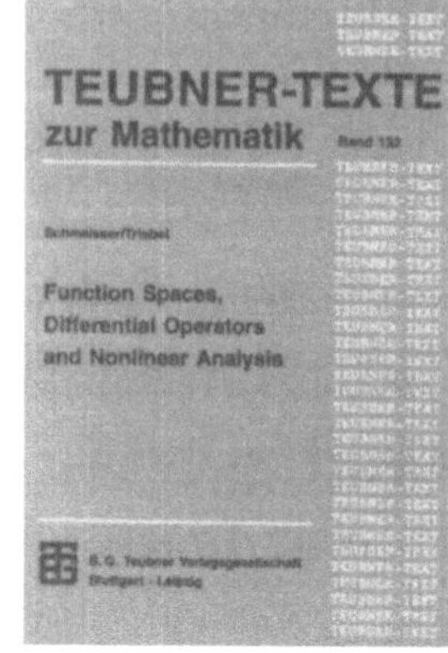

The book contains invited surveys and shorter communications presented at the International Conference "Function Spaces, Differential Operators and Nonlinear Analysis", which took place in Friedrichroda (Thuringia, Germany), autumn 1992. The main subjects and contributors are:
- Linear and nonlinear elliptic and parabolic boundary value problems (H. Amann, P. Drábek, S. I. Pohožaev, T. Runst)
- Functional calculus and singular operators in spaces of Besov and Sobolev type (G. Bourdaud, A. Youssfi),
- Hardy type operators in weighted functions spaces (M. Krbec, B. Opižc, L. Pick, J. Rákosnik).

Edited by
Hans-Jürgen Schmeisser and **Hans Triebel,** both Jena

1993. 308 pages.
16,2 x 23,5 cm.
Bound DM 54,80
ÖS 428,– / SFr 54,80
ISBN 3-8154-2045-8

(TEUBNER-TEXTE zur Mathematik, Vol. 133)

B. G. Teubner Verlagsgesellschaft
Stuttgart · Leipzig